This report contains the collective views of an international group of experts and does not necessarily represent the decisions or the stated policy of the United Nations Environment Programme, the International Labour Organisation, or the World Health Organization.

Environmental Health Criteria 117

METHYL ISOBUTYL KETONE

Published under the joint sponsorship of the United Nations Environment Programme, the International Labour Organisation, and the World Health Organization

First draft prepared by Dr K. Chipman, University of Birmingham, United Kingdom

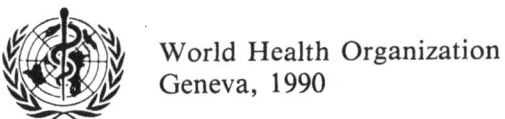

World Health Organization
Geneva, 1990

The **International Programme on Chemical Safety (IPCS)** is a joint venture of the United Nations Environment Programme, the International Labour Organisation, and the World Health Organization. The main objective of the IPCS is to carry out and disseminate evaluations of the effects of chemicals on human health and the quality of the environment. Supporting activities include the development of epidemiological, experimental laboratory, and risk-assessment methods that could produce internationally comparable results, and the development of manpower in the field of toxicology. Other activities carried out by the IPCS include the development of know-how for coping with chemical accidents, coordination of laboratory testing and epidemiological studies, and promotion of research on the mechanisms of the biological action of chemicals.

WHO Library Cataloguing in Publication Data

Methyl isobutyl ketone.

(Environmental health criteria ; 117)

1. Ketones - adverse effects   2. Ketones - toxicity   I. Series

ISBN 92 4 157117 9   (NLM Classification: QV 633)
ISSN 0250-863X

©World Health Organization 1990

Publications of the World Health Organization enjoy copyright protection in accordance with the provisions of Protocol 2 of the Universal Copyright Convention. For rights of reproduction or translation of WHO publications, in part or *in toto*, application should be made to the Office of Publications, World Health Organization, Geneva, Switzerland. The World Health Organization welcomes such applications.

The designations employed and the presentation of the material in this publication do not imply the expression of any opinion whatsoever on the part of the Secretariat of the World Health Organization concerning the legal status of any country, territory, city, or area or of its authorities, or concerning the delimitation of its frontiers or boundaries.

The mention of specific companies or of certain manufacturers' products does not imply that they are endorsed or recommended by the World Health Organization in preference to others of a similar nature that are not mentioned. Errors and omissions excepted, the names of proprietary products are distinguished by initial capital letters.

Printed in Finland
DHSS — Vammala — 5000

CONTENTS

ENVIRONMENTAL HEALTH CRITERIA FOR METHYL ISOBUTYL KETONE

1. SUMMARY 11
2. IDENTITY, PHYSICAL AND CHEMICAL PROPERTIES, ANALYTICAL METHODS 13
   2.1 Identity 13
   2.2 Physical and chemical properties 13
   2.3 Conversion factors 15
   2.4 Analytical methods 15
       2.4.1 Environmental media 15
           2.4.1.1 Air 15
           2.4.1.2 Water 16
           2.4.1.3 Tissues, body fluids, and skin washings 16
           2.4.1.4 Food 17
3. SOURCES OF HUMAN AND ENVIRONMENTAL EXPOSURE 18
   3.1 Natural occurrence 18
   3.2 Man-made sources 18
   3.3 Uses 18
4. ENVIRONMENTAL TRANSPORT, DISTRIBUTION, AND TRANSFORMATION 19
   4.1 Transport and distribution between media 19
   4.2 Biotransformation 19
       4.2.1 Biodegradation 19
       4.2.2 Abiotic degradation 19
       4.2.3 Photochemical smog reactivity 20
   4.3 Bioaccumulation 20
5. ENVIRONMENTAL LEVELS AND HUMAN EXPOSURE 21
   5.1 Environmental levels 21
       5.1.1 Air 21
       5.1.2 Food 21
       5.1.3 Water 21
       5.1.4 Soil 21
   5.2 Occupational exposure 22
       5.2.1 Exposure limit values 22

6.	KINETICS AND METABOLISM	25
	6.1 Experimental animals	25
	6.1.1 Effect on liver alcohol dehydrogenase *in vitro*	26
	6.2 Humans	26
7.	EFFECTS ON ORGANISMS IN THE ENVIRONMENT	28
	7.1 Microorganisms	28
	7.2 Aquatic organisms	28
	7.3 Terrestrial organisms	30
8.	EFFECTS ON EXPERIMENTAL ANIMALS AND *IN VITRO* TEST SYSTEMS	31
	8.1 Single exposures	31
	8.2 Short-term exposure	32
	8.2.1 Inhalation	32
	8.2.2 Oral	36
	8.2.3 Parenteral	37
	8.2.4 Skin application	37
	8.3 Skin, eye, and respiratory irritation; sensitization	38
	8.3.1 Skin irritation	38
	8.3.2 Eye irritation	38
	8.3.3 Respiratory irritation	38
	8.3.4 Skin sensitization	38
	8.4 Long-term exposure	38
	8.5 Reproduction, embryotoxicity, and teratogenicity	38
	8.6 Mutagenicity and related end-points	39
	8.6.1 Bacterial assays	39
	8.6.2 Yeast assay for mitotic gene conversions	41
	8.6.3 L5178Y TK+/- mouse lymphoma assay	42
	8.6.4 Unscheduled DNA synthesis in primary rat hepatocytes *in vitro*	42
	8.6.5 Mouse micronucleus assay	42
	8.6.6 Assay for structural chromosome damage	43
	8.6.7 Cell transformation assay	43
	8.7 Carcinogenicity	44
	8.8 Neurotoxicity	44
	8.9 *In vitro* toxicity assays	47

9. EFFECTS ON MAN 48

    9.1 Acute toxicity 48
    9.2 Short-term exposure 48
    9.3 Eye and respiratory irritation 48
    9.4 Long-term exposure 49
    9.5 Placental transfer 49
    9.6 Neurotoxicity 49

10. EVALUATION OF HUMAN HEALTH RISKS
    AND EFFECTS ON THE ENVIRONMENT 50

    10.1 Evaluation of effects on the environment 50
    10.2 Evaluation of health risks for humans 50

11. RECOMMENDATIONS 53

12. FURTHER RESEARCH 54

    REFERENCES 55

    RESUME 64

    EVALUATION DES RISQUES POUR
    LA SANTE HUMAINE ET DES EFFETS
    SUR L'ENVIRONNEMENT 67

    RECOMMANDATIONS 71

    RECHERCHES A EFFECTUER 72

    RESUMEN 73

    EVALUACION DE LOS RIESGOS PARA LA
    SALUD HUMANA Y DE LOS EFECTOS
    EN EL MEDIO AMBIENTE 75

    RECOMENDACIONES 78

    OTRAS INVESTIGACIONES 79

WHO TASK GROUP ON ENVIRONMENTAL HEALTH CRITERIA FOR METHYL ISOBUTYL KETONE (MIBK)

Participants

Professor E.A. Bababunmi, Department of Tropical Paediatrics, Liverpool School of Tropical Medicine, Liverpool, United Kingdom *(Rapporteur)*

Dr M. Cikrt, Centre of Industrial Hygiene and Occupational Diseases, Institute of Hygiene and Epidemiology, Prague, Czechoslovakia *(Vice-Chairman)*

Dr S. Dobson, Pollution and Ecotoxicology Section, Institute of Terrestrial Ecology, Monks Wood Experimental Station, Huntingdon, United Kingdom

Professor C.L. Galli, Toxicology Laboratory, Institute of Pharmacological Sciences, University of Milan, Milan, Italy *(Chairman)*

Dr S.D. Gangolli, British Industrial Biological Research Association, Carshalton, Surrey, United Kingdom

Dr C. Konantakieti, Technical Division, Food and Drug Administration, Ministry of Public Health, Bangkok, Thailand

Dr O. Ladefoged, Laboratory of Pathology, Institute of Toxicology, National Food Agency of Denmark, Ministry of Health, Soborg, Denmark

Professor A. Massoud, Department of Community Environmental and Occupational Medicine, Ainshams Faculty of Medicine, Cairo, Egypt

Dr V. Riihimäki, Department of Industrial Hygiene and Toxicology, Institute of Occupational Health, Helsinki, Finland

Secretariat

Dr P.G. Jenkins, International Programme on Chemical Safety, World Health Organization, Geneva, Switzerland

Ms B. Labarthe, International Register of Potentially
   Toxic Chemicals, United Nations Environment Programme,
   Geneva, Switzerland

Dr E. Smith, International Programme on Chemical Safety,
   World Health Organization, Geneva, Switzerland
   (Secretary)

NOTE TO READERS OF THE CRITERIA DOCUMENTS

Every effort has been made to present information in the criteria documents as accurately as possible without unduly delaying their publication. In the interest of all users of the environmental health criteria documents, readers are kindly requested to communicate any errors that may have occurred to the Manager of the International Programme on Chemical Safety, World Health Organization, Geneva, Switzerland, in order that they may be included in corrigenda, which will appear in subsequent volumes.

* * *

A detailed data profile and a legal file can be obtained from the International Register of Potentially Toxic Chemicals, Palais des Nations, 1211 Geneva 10, Switzerland (Telephone No. 7988400 or 7985850)

ENVIRONMENTAL HEALTH CRITERIA FOR METHYL ISOBUTYL KETONE

A WHO Task Group on Environmental Health Criteria for Methyl Isobutyl Ketone met at Carshalton, United Kingdom, from 12 to 16 March 1990. Dr E.M. Smith, IPCS, opened the meeting on behalf of the heads of the three IPCS cooperating organizations (UNEP/ILO/WHO). The Task Group reviewed and revised the draft monograph and made an evaluation of the health risks of exposure to methyl isobutyl ketone.

The first draft of this document was prepared by Dr K. Chipman, University of Birmingham, United Kingdom. The second draft was also prepared by Dr Chipman following circulation of the first draft to IPCS contact points for Environmental Health Criteria monographs. Particularly valuable comments on the draft were made by the United Kingdom Department of Health, the European Chemical Industry Ecology and Toxicology Centre (ECETOC), and the US Environmental Protection Agency, National Institute of Environmental Health Sciences, and National Institute of Occupational Safety and Health.

Dr E.M. Smith and Dr P.G. Jenkins, both members of the IPCS Central Unit, were responsible for the technical development and editing, respectively, of this monograph.

The efforts of all who helped in the preparation and finalization of the document are gratefully acknowledged.

* * *

Financial support for this Task Group was provided by the United Kingdom Department of Health as part of its contributions to IPCS.

Partial financial support for the publication of this criteria document was kindly provided by the United States Department of Health and Human Services, through a contract from the National Institute of Environmental Health Sciences, Research Triangle Park, North Carolina, USA - a WHO Collaborating Centre for Environmental Health Effects.

ABBREVIATIONS

CLV	ceiling value
GC	gas chromatography
GRAS	generally regarded as safe
HMP	4-hydroxy-4-methyl-2-pentanone
ip	intraperitoneal
MIBK	methyl isobutyl ketone
NOEL	no-observed-effect level
QSAR	quantitative structure activity relationship
TLV	threshold limit value

1. SUMMARY

Methyl isobutyl ketone (MIBK) is a clear liquid with a sweet odour and is produced commercially for wide use as a solvent. It can be measured by gas chromatography with flame ionization detection. It rapidly evaporates into the atmosphere, where it is rapidly phototransformed. MIBK is readily biodegradable, and this, together with its moderate water solubility and low octanol/water partition coefficient, suggests that it has a low bioaccumulation potential. Occupational exposure limits range from 100-410 mg/m^3 (time-weighted average, TWA) and 5-300 mg/m^3 (ceiling value, CLV) in different countries.

MIBK is readily metabolized to water-soluble excretory products and is of low acute systemic toxicity in animals by the oral and inhalation routes of exposure. Peripheral axonopathy has not been reported in animal studies. There are no accurate LC_{50} data. A 4-h exposure to 16 400 mg/m^3 (4000 ppm) was lethal in rats. Liquid MIBK and vapour concentrations in the range 10-410 mg/m^3 (2.4-100 ppm) are irritant to the eyes and the upper respiratory tract. Concentrations up to 200 mg/m^3 (50 ppm) produced no significant effects on humans in a simple reaction time task or a test of mental arithmetic. Prolonged or repeated skin contact may cause drying and flaking of the skin. Accidental aspiration of liquid MIBK can cause chemical pneumonitis.

In a 90-day gavage study on rats, a NOEL of 50 mg/kg per day was found. In 90-day inhalation studies on rats and mice, concentrations of up to 4100 mg/m^3 (1000 ppm) did not result in any life-threatening signs of toxicity. However, compound-related reversible morphological changes in the liver and kidney were reported. In a number of studies, MIBK concentrations as low as 1025 mg/m^3 (250 ppm) were capable of increasing liver size. With exposure to 4100 mg/m^3 (1000 ppm) for 50 days, microsomal enzyme metabolism was induced in the livers of chickens. Effects at higher doses (up to 8180 mg/m^3, 1996 ppm) were limited to increased liver weight with no histological damage. In 90-day studies with mice, rats, dogs, and monkeys, only male rats developed hyaline droplets in the

Summary

proximal tubules of the kidney (hyaline droplet toxic tubular nephrosis). This effect in male rats was reversible and of doubtful significance for humans. Enzyme induction may be the basis of potentiation of haloalkane toxicity by MIBK. MIBK was also able to potentiate the cholestatic effect of manganese given with or without bilirubin.

In baboons exposed for 7 days to 205 mg/m^3 (50 ppm), effects on neurobehaviour were reported.

MIBK is fetotoxic at a concentration that produces definite maternal toxicity (12 300 mg/m^3, 3000 ppm) but is not embryotoxic or teratogenic at this concentration. At a concentration of 4100 mg/m^3 (1000 ppm), it was neither embryotoxic, fetotoxic, nor teratogenic in rats or mice.

MIBK has been studied for genotoxicity in a number of short-term assays, including *in vitro* bacterial, yeast, and mammalian cell tests and a micronucleus assay in mice. These studies indicate that MIBK is not genotoxic. No reports of long-term or carcinogenicity studies are available.

At 410 mg/m^3 (100 ppm) MIBK can induce in humans symptoms such as eye irritation, headaches, nausea, dizziness, and fatigue consistent with a reversible depressant effect on the central nervous system, but there is no evidence that it produces permanent damage to the nervous system.

MIBK has low toxicity for aquatic organisms and microorganisms.

The relatively high volatility, rapid atmospheric phototransformation, ready biodegradability, and low mammalian and aquatic toxicity of MIBK indicate that adverse environmental effects of this substance are only likely to occur after accidental spills or from uncontrolled industrial effluents.

2. IDENTITY, PHYSICAL AND CHEMICAL PROPERTIES, ANALYTICAL METHODS

2.1 Identity

Common name: Methyl isobutyl ketone (MIBK)

Chemical structure:

$$H_3C-\underset{\|}{\overset{O}{C}}-\underset{|}{\overset{H}{\underset{H}{C}}}-\underset{|}{\overset{CH_3}{\underset{H}{C}}}-CH_3$$

Chemical formula: $C_6H_{12}O$

Relative molecular mass: 100.16

Common synonyms: MIBK, MIK, 4-methyl-2-pentanone, 2-methyl-4-pentanone, hexanone, hexone, isopropyl-acetone, 4-methyl pentan-2-one, 4-methyl-2-oxopentane, 2-methyl propyl methyl ketone, isobutylmethyl ketone

CAS registry number: 108-10-1

A typical sample of MIBK has a purity of 99% (by mass); it may contain the following impurities: dimethyl heptane (< 0.3%), water (< 0.1%), methyl isobutyl carbinol (< 0.06%), mesityloxide (< 0.03%), acidity as acetic acid (< 0.002%), and non-volatiles (< 0.002%).

2.2 Physical and chemical properties

MIBK is a clear liquid with a sweet odour.

Some physical and chemical properties of MIBK are given in Table 1. The partition coefficients of MIBK are 79 for water/air, 90 for blood/air, and 926 for oil/air

(Sato & Nakajima, 1979). MIBK can react violently with oxidizing agents such as peroxides, nitrates, and perchlorates, reducing agents, or with potassium *tert*-butoxide. When heated, MIBK may form peroxides by auto-oxidation and these may explode spontaneously (Sax, 1979).

Table 1. Some physical and chemical properties of MIBK[a]

Property	Value
Physical form	liquid
Colour	colourless
Odour/taste	sweet
Odour threshold limit (mg/m^3)	1.64 (0.4 ppm)[b] 1.68 (0.41 ppm)[a]
Boiling point (°C at 101 KPa)	116.2 (range, 116 to 119)[c]
Freezing point (°C)	-80.26 (range, -80 to -85)[c]
Specific gravity (20°C/4°C)	0.8017
Refractive index (n_D^{20})	1.395 to 1.397
Viscosity (mPa·s) (20°C)	0.58 to 0.61
Vapour density (air = 1)	3.45
Vapour pressure (KPa) (20°C)	1.99
Concentration in saturated air (g/m^3) (20°C) (101 KPa)	27
Flashpoint (°C) (closed cup)	14
Auto-ignition temperature (°C)	460
Explosion limits in air (101 KPa) (% vol.)	1.4 to 7.5
Solubility in water (20°C) (g/litre)	17
Octanol/water partition coefficient (log P_{ow})	1.38[d]

[a] From: Verschueren (1983).
[b] From: Ruth (1986).
[c] For commercial products.
[d] From: Leo & Weininger (1984).

2.3 Conversion factors

1 ppm = 4.1 mg/m^3
1 mg/m^3 = 0.244 ppm

2.4 Analytical methods

2.4.1 Environmental media

Gas chromatography (GC) is suitable for analysing trace quantities of MIBK (Analytical Quality Control, 1972; Webb et al., 1973) and the use of fused capillary columns is advantageous. The use of bonded-phase capillary columns may overcome the need for solid or liquid phase extraction of samples. Flame ionization detection (FID) is very sensitive (Webb et al., 1973), while mass spectroscopy is particularly useful for identifying MIBK in complex media.

2.4.1.1 Air

Measurement of MIBK in air involves sampling (10-12 litres at a rate of 0.2 litre/min) on charcoal, silica gel, or some chromatographic column packings, followed by desorption with carbon disulfide and further analysis by gas chromatography with flame ionization detection (Tomczyk & Rogaczewska, 1979; NIOSH, 1984; Moshlakova & Indina, 1986). The method has been validated over the range of 208-836 mg/m^3 (52-209 ppm) and the probable useful concentration range is 40-1230 mg/m^3 (10-300 ppm) (NIOSH, 1984). No interferences were reported. Using this technique, Bamberger et al. (1978) reported an average desorption efficiency of approximately 81% following the exposure of a dosimeter for 5 h to 103 or 410 or 820 mg MIBK/m^3 (25 or 100 or 200 ppm). Storage in a covered dosimeter for 2 weeks reduced the recovery to 69%. Levin & Carleborg (1987) investigated a range of adsorbents for work-room air sampling of MIBK. The retention capacity was as follows (for 5 mg generated at 85% relative humidity at 0.2 litres/min in 5 litres air): XAD-2, 21%; XAD-4, 65%; XAD-7, 39%; activated charcoal, 97%; Ambersorb XE-348, 98%. Although recovery was reduced to 61% following storage of samples on charcoal, recoveries were not reduced under these conditions for Ambersorb XE-348.

MIBK can also be sampled efficiently by the use of Tenax-GC®, a polymer of 2,6-diphenyl-*p*-phenylene oxide. Its main advantages are its high temperature stability and low affinity for water vapour. For MIBK, collection efficiency is 99% using 135 g Tenax-GC®/17 litres of air. This method is more sensitive than the charcoal/solvent desorption technique and produces samples that remain stable for 6 months (Brown & Purnell, 1979).

2.4.1.2 Water

Techniques such as head-space sampling when there is no interference (Corwin, 1969), liquid-liquid extraction (Keith, 1974; Austern et al., 1975), distillation, or stripping with an inert gas stream (Webb et al., 1973; Ellison & Wallbank, 1974) have been used, because water is not a suitable solvent for gas chromatographic analysis. The use of synthetic resin gas chromatographic columns gives low detection limits (μg/ml range) and high recovery; for example, they have been used in the analysis of traces of MIBK in drinking-water (Burnham et al., 1972). Ellison & Wallbank (1974) removed MIBK from waste water and waste sludges by steam distillation and partitioning into cyclohexanone before gas chromatographic analysis.

2.4.1.3 Tissues, body fluids, and skin washings

The presence of MIBK and other 2-pentanones in 24-h urine samples of unexposed human beings has been demonstrated using gas chromatography and mass spectrometry (Zlatkis & Liebich, 1971). Bellanca et al. (1982) used both gas chromatography and mass spectroscopy in the electron ionization mode and, for more sensitive detection, employed select ion monitoring coupled with capillary gas chromatography. These methods were used to detect MIBK liberated into head-space gas from samples of brain, liver, lung, vitreous fluid, kidney, and blood. MIBK and its metabolites have been detected by gas chromatography in the serum of guinea-pigs administered MIBK (DiVincenzo et al., 1976), and Moshlakova & Indina (1986) have detected MIBK (0.006 to 0.06 mg/cm^2) in skin washings of workers using gas chromatography and flame ionization detection.

2.4.1.4 Food

Residual MIBK in food packaging films can be analysed by gas chromatography (Raccio & Widomski, 1981; Fernandes, 1985). For milk analyses, gas chromatography can be combined with mass spectroscopy (Weller & Wolf, 1989).

3. SOURCES OF HUMAN AND ENVIRONMENTAL EXPOSURE

3.1 Natural occurrence

MIBK occurs naturally in food, is a permitted flavouring agent with GRAS status in the USA, and is used in food contact packaging materials. It is found in a wide range of foods, e.g., fruits, baked potatoes, cheese, milk, some meats, and some alcoholic beverages. The following values have been reported: papaya, 8 µg/kg; beer, 10 to 120 µg/kg; coffee, 6.5 mg/kg (TNO 1983a,b; 1986; 1987).

3.2 Man-made sources

MIBK is produced commercially by acetone condensation, followed by catalytic hydrogenation in a one-step catalytic process. The annual production in the USA in 1975 was estimated to be 80 500 tonnes (Lande et al., 1976), and the annual global production in 1975 was estimated to be 250 000 tonnes (OECD, 1977). Consumption in the European Economic Community (EEC) in 1981 was estimated to be 45 000 tonnes (ECDIN, 1990).

3.3 Uses

MIBK is used as one of the component ketones in lacquers, such as cellulose and polyurethane lacquers (Sabroe & Olsen, 1979), and as a minor component of paint solvents, including car and industrial spray paints (Hänninen et al., 1976; Elofsson et al., 1980). It also has uses as an extraction solvent, e.g., in pharmaceutical products and in the manufacture of methyl amyl alcohol and as a denaturant for ethyl alcohol (Zakhari et al., 1977).

4. ENVIRONMENTAL TRANSPORT, DISTRIBUTION, AND TRANSFORMATION

4.1 Transport and distribution between media

There are no experimental data on the transport, mobility, and concentration of MIBK in the environment.

MIBK is moderately soluble in water and volatilizes only slowly from soil and surface waters. A theoretical half-life of 33 days, in a body of water with a depth of 1 m, can be calculated according to the Fugacity Model of MacKay & Wolkoff (1973). Based on its moderate water solubility and low soil adsorption coefficient, MIBK is a potential contaminant of ground water (see section 5.1.3).

4.2 Biotransformation

4.2.1 Biodegradation

Using the standard dilution method with sludge from a waste-treatment plant, Bridié et al. (1979b) found a biological oxygen demand after 5 days at 20°C (BOD_5) for MIBK of 76% of the theoretical oxygen demand (ThOD). MITI (1978) confirmed that MIBK was readily biodegradable in fresh water and sea water. Price et al. (1974) studied the biodegradability of MIBK and found that the non-acclimated extent of bio-oxidation was 56, 66, 69, and 69% at 5, 10, 15, and 20 days, respectively, in fresh water. The respective values in synthetic salt water were 15, 46, 50, and 53%. The measured chemical oxygen demand (COD) was 2.4 mg/mg.

Data are not available on the biodegradation of MIBK in soil.

4.2.2 Abiotic degradation

MIBK is degraded in the atmosphere by OH radicals. Cox et al. (1980) and Atkinson et al. (1982) found k_{OH} reactivity constants of 12.4×10^{-12} and 14.5×10^{-12} cm^3/mol per sec, respectively, which corresponded to half-lives of 0.57 and 0.55 days. MIBK is also photodegraded. The major phototransformation product is acetone, which has a k_{OH} of 0.5×10^{-12} cm³/mol per sec, corresponding to a half-life of 16 days (Cox et al., 1980).

4.2.3 Photochemical smog reactivity

There is some experimental evidence indicating the participation of ketones in the photochemical smog cycle as free-radical chain initiators. However, their contribution to overall smog generation has not been established, but is thought to be minor (Lande et al., 1976).

4.3 Bioaccumulation

There are no data on the ability of MIBK to accumulate in biological material. However, its moderate water solubility and low octanol/water partition coefficient (log P_{ow}) suggest that it has low bioaccumulation potential (OECD, 1984). MIBK is not expected to be persistent. It will probably volatilize fairly readily except in wet environments and may be oxidized in the atmosphere. Due to its low log P_{ow}, it is unlikely that it will be appreciably absorbed.

5. ENVIRONMENTAL LEVELS AND HUMAN EXPOSURE

5.1 Environmental levels

5.1.1 Air

MIBK release into the atmosphere may occur during its production through fugitive emissions and incomplete removal of vapours from reaction gases before they are vented or disposed of in a scrubber. In the Federal Republic of Germany, MIBK belongs to class III, air emissions of which must not exceed (as the sum of all compounds in any class) 150 mg/m^3 (37 ppm) at a mass flow of 3 kg/h or more. The maximum recommended ambient concentrations are 0.2 mg/m^3 (0.05 ppm) in Czechoslovakia and must not exceed 0.1 mg/m^3 (0.025 ppm) in the USSR (IRPTC, 1990). MIBK has been detected in automotive exhaust emissions (Hampton et al., 1982).

5.1.2 Food

MIBK is allowed as a component of food packaging and is allowed to come in contact with food materials in the USA and the EEC. The EEC limit for the *sum* of all permitted solvents is 60 mg/m^2 on the side with food contact. The intake via food flavourings based on a 1970 survey of usage in the USA was estimated to be 3.35 mg/person per day (NTIS, 1985). The levels in particular foods were: baked goods, 10.9 mg/kg; frozen dairy products 11.5 mg/kg; meat products, 2.6 mg/kg; soft candy, 12.3 mg/kg; gelatins, puddings, 10.9 mg/kg; beverages, 10.2 mg/kg.

MIBK has also been detected in human breast milk (Pellizzari et al., 1982).

5.1.3 Water

MIBK may be released during the discharge of spent scrubbing water from industrial production processes. Traces of MIBK have been found in tap water in the USA (CEC, 1976) and in the United Kingdom (Fawell & Hunt, 1981). MIBK is included in the Code of Federal Regulations

(CFR, 1987), which lays down methods for the analysis of organic chemicals in ground water at hazardous waste sites. MIBK has frequently been detected in leachates from waste sites (Francis et al., 1980; Sawhney & Kozloski, 1984; Garman et al., 1987; Brown & Donnelly, 1988).

5.1.4 Soil

MIBK can contaminate soil as a result of accidental spillage or disposal of solid wastes or sludges (Basu et al., 1968), but there are no data on levels in soil.

5.2 Occupational exposure

Closed production systems should ensure that occupational exposures are below recommended occupational exposure limits. However, emissions that occur when MIBK is used as a solvent, e.g., in paints and lacquers, are less easily controlled. Hänninen et al. (1976) reported a mean time-weighted average (TWA) concentration of 7 mg/m^3 (range, 4-160 mg/m^3) (1.7 ppm; range, 1-39 ppm) in the breathing zone of spray painters in car repair shops. Residual MIBK contained in plastic products can outgas under reduced pressure conditions and may appear as a contaminant in the environment of spacecraft (MacEwen et al., 1971). It has been detected at levels of < 0.005 to 0.02 mg/m^3 in the atmosphere of spacecraft (Rippstein & Coleman, 1984). MIBK has also been found as a volatile degradation product of polypropylene at temperatures of 220 or 280 °C (Frostling et al., 1984). According to a study by Kristensson & Beving (1987), exposure measurements for workers painting indoors for periods of 6-8 h indicated that the concentrations of probe solvents (including MIBK) were usually well below prescribed threshold limit values.

5.2.1 Exposure limit values

Exposure limit values for various countries are given in Table 2. The USSR requires special skin and eye protection for workers exposed to MIBK.

Table 2. Some national occupational air exposure limits
used in various countries[a]

Country/ organization	Exposure limit description[b]	Value (mg/m³)	Effective date
Australia	Recommended threshold limit value (TLV) - Time-weighted average (TWA) - Short-term exposure limit (STEL)	205 300	1985(r)
Belgium	Recommended threshold limit value (TLV) - Time-weighted average (TWA) - Short-term exposure limit (STEL)	205 300	1988(r)
Finland	Occupational exposure limit (MPC) - Time-weighted average (TWA) - Short-term exposure limit (STEL)	210 315	1987
Germany, Federal Republic of	Recommended threshold limit value (MAK) - Time-weighted average (TWA) - Short-term exposure limit (STEL)	400 2000	1988(r)
Japan	Administrative concentration - Time-weghted average (TWA)	205	1990(n)
Netherlands	Recommended threshold limit value (MXL) - Time-weighted average (TWA)	240	1989(r)
Poland	Permissible exposure limit (MPC) - Time-weighted average (TWA)	200	1982(r)
Romania	Permissible exposure limit (MPC) - Time-weighted average (TWA) - Ceiling value (CLV)	200 300	1984(r)
Switzerland	Permissible exposure limit (MAK) - Time-weighted average (TWA)	205	1987(r)
Sweden	Permissible exposure limit (HLV) - Time-weighted average (TWA) - Short-term exposure limit (STEL)	100 200	1990(n)
United Kingdom	Occupational exposure standard (OES) - Time-weighted average (TWA) - Short-term exposure limit (STEL)	205 300	1990(n)
USA (ACGIH)	Recommended threshold limit value (TLV) - Time-weighted average (TWA) - Short-term exposure limit (STEL)	205 300	1987(r)
(OSHA)	Permissible exposure limit (PEL) - Time-weighted average (TWA) - Short-term exposure limit (STEL)	205 300	1990(n)

Environmental Levels and Human Exposure

Table 2 (contd).

Country/ organization	Exposure limit description[b]	Value (mg/m^3)	Effective Date
USSR	Temporary exposure limit (TSEL) - Ceiling value (CLV)	5	1989
Yugoslavia	Permissible exposure limit (MAC) - Time-weighted average (TWA)	410	1971(r)

[a] From IRPTC, 1990
[b] TWA = a maximum mean exposure limit based generally over the period of a working day.
STEL = a maximum concerntration of exposure for a specified time duration (generally 10-30 min.)
(n) = directly notified by countries
Where no effective date appears in the IRPTC legal file, the year of the reference from which the data are taken is shown, indicated by (r).

6. KINETICS AND METABOLISM

6.1 Experimental animals

Following the intraperitoneal (ip) injection of 450 mg MIBK/kg body weight to guinea-pigs, two metabolites were found in the serum (DiVincenzo et al., 1976). The major metabolite, 4-hydroxy-4-methyl-2-pentanone (HMP), was formed by oxidation of MIBK, while a minor metabolite, 4-methyl-2-pentanol, was formed by reduction of MIBK (Fig. 1). The serum half-life and total clearance time for parent MIBK were calculated as 66 min and 6 h, respectively, whereas 4-hydroxy-4-methyl-2-pentanone (HMP) was cleared in 16 h. The hydroxylation products of MIBK, such as 4-methyl-2-pentanol, are expected either to be conjugated with sulfate or glucuronic acid and excreted in the urine or to enter intermediary metabolism to be converted to carbon dioxide.

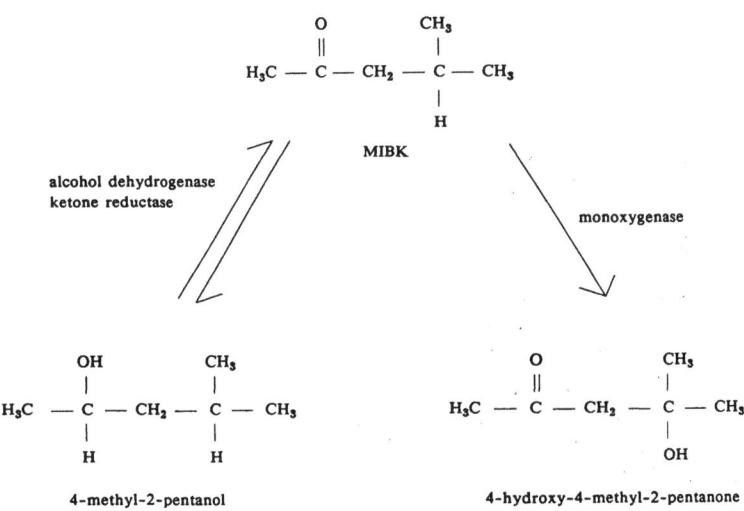

Fig. 1. Metabolic pathway of MIBK in the guinea-pig and rat
From: DiVincenzo et al. (1976); Pilon (1987).

HMP and 4-methyl-2-pentanone have also been identified in rats (Pilon, 1987). MIBK increased aniline hydroxylase activity and cytochrome P-450 concentration in chicken liver microsomes after inhalation (4100 mg/m^3, 1000 ppm, for 50 days) (Abou-Donia et al., 1985b). The induction by MIBK was apparently of similar capacity to that of methyl n-butyl ketone (Abou-Donia et al., 1985a). It is likely that MIBK can induce its own oxidative metabolism as well as that of other substances. The study of Malyscheva (1988) (section 8.2.4) suggested that MIBK is absorbed through the skin as well as via the oral and inhalation routes (section 8.1). A comparison of intraperitoneal and oral LD_{50} values (Zakhari et al., 1977) suggests an oral absorption of 30% or more. Malyscheva (1988) showed dermal absorption followed by extensive distribution (toxic signs in many organs). Likewise, inhalation studies producing liver and kidney changes are suggestive of extensive distribution (MacEwen et al., 1971; Vernot et al., 1971).

The structure of MIBK precludes the metabolic production of 2,5-hexane-dione, the neurotoxic agent formed from both hexane and methyl n-butyl ketone.

6.1.1 Effect on liver alcohol dehydrogenase in vitro

MIBK in N,N-dimethylacetamide has been shown to reduce the activity of mouse liver alcohol dehydrogenase *in vitro* (Cunningham et al., 1989).

6.2 Humans

MIBK and other substituted 2-pentanones have been reported in urine samples from unexposed humans (Zlatkis & Liebich, 1971). MIBK was detected in various tissues and body fluids (section 2.4.1.3) of two individuals who suffered fatal exposure to a mixture of organic solvents. The concentration range of MIBK in the tissues and body fluids of the two decedents was 1.4-2.5 and 0.2-0.8 mg/kg, respectively. The tissue distribution differed markedly between the two individuals (Bellanca et al., 1982). The inhalation study carried out by Dick et al. (1990) with human volunteers suggested that exposures to 410 mg MIBK/m^3 (100 ppm) for 4 h causes steady state blood levels to be attained. Blood and breath samples collected

90 min after exposure indicated essentially complete clearance of the absorbed MIBK. Hjelm et al. (1990) exposed human volunteers for 2 h during light physical excercise to MIBK (10, 100, and 200 mg/m^3; 2.4, 24.4, and 48.8 ppm). The concentration of MIBK in blood rose rapidly after the onset of exposure, during which no plateau level was reached. No tendency towards saturation kinetics was observed over the dose range, the apparent blood clearance being 1.6 litres/h per kg throughout. Only 0.04% of the total MIBK dose was eliminated unchanged via the kidneys within 3 h after exposure.

7. EFFECTS ON ORGANISMS IN THE ENVIRONMENT

Toxicity data reported in this chapter should be interpreted with caution, since the tests were conducted under static conditions using nominal rather than measured concentrations. Actual concentrations experienced by the test organisms cannot be determined for these tests.

7.1 Microorganisms

MIBK has low toxicity for microorganisms as indicated by the threshold concentrations required for inhibition of growth (Table 3).

Table 3. Toxicity of MIBK for microorganisms

Species	Threshold concentration for growth inhibition (mg/litre)	Duration of study	Reference
Protozoa			
Saprozoic flagellate (*Chilomonas paramaecium*)	> 800	48 h	Bringmann & Kühn (1981)
Bacteriovorous flagellate (*Entosiphon sulcatum*)	450	72 h	Bringmann & Kühn (1981)
Bacteriovorous ciliate (*Uronema parduczi*)	950	20 h	Bringmann & Kühn (1981)
Bacterium			
Pseudomonas putida	275	16 h	Bringmann & Kühn (1977b)

7.2 Aquatic organisms

MIBK appears to have low toxicity for aquatic organisms (Table 4). The maximum zero lethality concentration (LC_0) is in the range of 480-720 mg/litre (Juhnke &

Table 4. Acute toxicity of MIBK for aquatic organisms

Species	LC$_{50}$ (mg/litre)	Duration of study	Reference
Freshwater fish			
Golden orfe (Leuciscus idus melanotus)	672-744	48 h	Juhnke & Lüdemann (1978)
Goldfish (Carassius auratus)	460	24 h	Bridié et al. (1979b)
Fathead minnow (Pimephales promelas)	≈525	96 h	Call et al. (1985)
Invertebrates			
Freshwater			
Water flea (Daphnia magna)	4280 1550	24 h 24 h	Bringmann & Kühn (1977a) Bringmann & Kühn (1982)
Marine			
Brine shrimp (Artemia salina)	1230	24 h	Price et al. (1974)
Freshwater algae			
Green algae (Scenedesmus quadricauda)	725[a]	8 days	Bringmann & Kühn (1977b)
Bluegreen algae (Microcystis aeruginosa)	136[a]	8 days	Bringmann & Kühn (1978)

[a] Threshold concentration for reduction of total biomass

Lüdemann, 1978). Using a QSAR model, Lipnick et al. (1987) showed that the 24-h LC$_{50}$ of 460 mg/litre reported by Bridié et al. (1979a) in goldfish *(Carassius auratus)* fitted a narcotic mechanism of action.

Aquatic invertebrates are less sensitive than fish to the toxicity of MIBK, 24-h LC$_{50}$ values of 4280 mg/litre (Bringmann & Kuhn, 1977a) and 1550 mg/litre (Bringmann & Kuhn, 1982) having been reported for the water flea *Daphnia magna* and 1230 mg/litre for the brine shrimp *Artemia salina* (Price et al., 1974). The LC$_{100}$ and LC$_0$ values for

Daphnia magna were 5000 and 2280 mg/litre, respectively (Bringmann & Kuhn, 1977a).

The toxicity of MIBK was also measured in the green alga *Scenedesmus quadricauda*, in which the 8-day threshold for toxicity was 725 mg/litre (Bringmann & Kuhn, 1977b), and in the relatively more sensitive cyanobacterium (blue-green alga) *Microcystis aeruginosa*, in which the toxicity threshold was 136 mg/litre (Bringmann & Kuhn, 1978).

7.3 Terrestrial organisms

Experimental data are not available.

8. EFFECTS ON EXPERIMENTAL ANIMALS AND *IN VITRO* TEST SYSTEMS

8.1 Single exposures

MIBK is of low acute toxicity by the oral and inhalation routes of exposure (Table 5).

The maximum time for which rats could be exposed to a saturated atmosphere of MIBK without dying was 15 min (Smyth et al., 1951). In one study, six rats survived a 4-h exposure to 8200 mg MIBK/m^3 (2000 ppm), but, following a 4-h exposure to 16 400 mg/m^3 (4000 ppm), all six animals died within 14 days.

In studies by Specht (1938) and Specht et al. (1940), female guinea-pigs were exposed to MIBK concentrations of 4100, 69 000, and 115 000 mg/m^3 (1000, 16 800, and 28 000 ppm, respectively) for up to 24 h. In view of the method used for generating the atmosphere (allowing measured amounts of MIBK to evaporate freely to one cubic meter volume of air at 25-26 °C), the two higher levels must be greatly exaggerated because the saturation concentration in air for MIBK at 25 °C is 40 000 mg/m^3. At 4100 mg/m^3 there was minimal eye or nasal irritation. However, there was a decreased respiratory rate during the first 6 h of exposure, which was attributed to a narcotic effect. The higher levels produced obvious signs of eye and nose irritation, followed by salivation, lacrimation, ataxia, progressive narcosis, and death. The highest level killed 50% of the animals within 45 min. Autopsy and histopathological investigations in some animals showed fatty livers and congestion of the brain, lungs, and spleen, but no damage to the heart and kidneys was observed.

A single ip injection of 500 mg MIBK/kg body weight in guinea-pigs did not induce changes in the serum ornithine-carbamyl transferase level, and there were no histopathological changes in the liver. However, an injection of 1000 mg/kg body weight killed one out of four animals, and a slight increase in the serum ornithine-carbamyl transferase level was seen in the survivors 24 h after dosing.

Histopathologically, there was possible lipid accumulation in liver cells but no evidence of liver damage (Divincenzo & Krasavage, 1974). It should be noted that an ip injection of 560 mg/kg produced minimal effects in rats (Vezina et al., 1985), suggesting that mice are more sensitive than rats (the ip LD_{50} is 590 mg/kg in mice, see Table 5).

In male Sprague-Dawley rats, a single oral dose of MIBK enhanced the hepatotoxicity of a single ip dose of chloroform given 24 h later. The no-observed-effect and minimal-effect levels of MIBK were 375 and 560 mg/kg body weight, respectively (Vezina et al., 1985). The ketone potentiation of haloalkane-induced hepatonecrosis has been attributed to enhanced bioactivation of the haloalkane, which is mediated by the increased cytochrome P-450 activity induced by the ketone (Branchflower et al., 1983). The extent of potentiation of carbon tetrachloride liver toxicity (as shown by an increase in plasma alanine transaminase activity and bilirubin concentration) was found to depend on the concentration of both MIBK and carbon tetrachloride in male rats. The minimum effective MIBK dose decreased 10-fold when the carbon tetrachloride dose was increased from 0.01 ml/kg to 0.1 ml/kg. These findings suggest that liver injury is determined by the product of MIBK and carbon tetrachloride doses (Pilon et al., 1988). Attention should be paid to this when working in environments containing mixtures of solvents.

8.2 Short-term exposure

8.2.1 Inhalation

In rats exposed to 410 mg/m^3 (100 ppm) for 2 weeks, there was an increase in kidney weight. An increase in both liver and kidney weights was observed after exposure to 820 mg/m^3 (200 ppm) for 2 weeks or 410 mg/m^3 (100 ppm) for 90 days. Histopathological investigations showed hyaline droplets in the proximal tubules of the kidney and this finding was named hyaline droplet toxic tubular nephrosis (MacEwen et al., 1971; Vernot et al., 1971). In rats, dogs, and monkeys exposed continuously for 90 days to MIBK at 410 mg/m^3 (100 ppm) under reduced oxygen tension and reduced atmospheric pressure (65% oxygen at

Table 5. LD$_{50}$ and LC$_{50}$ values for MIBK in rats and mice

Route of exposure	Species	Duration of exposure	LD$_{50}$ or LC$_{50}$	Reference
Oral (LD$_{50}$)	rat		4570 mg/kg body weight	Smyth et al. (1951)
	rat		4600 mg/kg body weight	Batyrova (1973)
	rat		2080 mg/kg body weight	RTECS (1987)
	mouse		2850 mg/kg body weight	Batyrova (1973)
	mouse		1900 mg/kg body weight	Zakhari et al. (1977)
Intraperitoneal (LD$_{50}$)	mouse		590 mg/kg body weight	Zakhari et al. (1977)
Inhalation (LC$_{50}$)	rat	4 h	8.2–16.4 g/m^3	Smyth et al. (1951); Smyth (1956)
	mouse	2 h	20.5 g/m^3	Batyrova (1973)
	mouse	45 mins	74.2 g/m^3 (18 105 ppm)	Zakhari et al. (1977)

34.7 k Pa), liver and kidney weights were increased in rats after 90 days. Hyaline droplets were observed in rat kidney epithelium after 15 days, but this effect was reversible after a 3- to 4-week recovery period. No histopathological changes were reported in monkeys or dogs and no sex differences were reported for any parameter (MacEwen et al., 1971).

Four groups of six male and six female F-344 rats and six male and six female B6C3F$_1$ mice were exposed to 0, 44, 2050, or 8180 mg MIBK/m^3 (0, 10.1, 501, or 1996 ppm, respectively) for 6 h/day (for 5 days with 2 days off followed by 4 more consecutive days of exposure) (Dodd et al., 1982; Phillips et al., 1987). Lacrimation was observed in the highest dose group, but no ophthalmological lesions or alterations in body weight gain were found. Liver weight, expressed as a percentage of body weight, was increased in male and female rats and in female mice exposed to 8180 mg/m^3. A statistically significant increase in liver weight was also observed in male rats exposed to 2050 mg/m^3. Male and female rats and female mice showed a significant increase in both absolute and relative kidney weights when exposed to 8180 mg/m^3. However, male mice at this exposure level exhibited a significant decrease in relative kidney weight. Of the animals exposed to 2050 mg/m^3, only male rats exhibited an increase in kidney weight, but this was not statistically significantly different from the control value. Hyaline droplet formation was seen in the kidneys of male rats exposed to 2050 and 8180 mg/m^3. Epithelial regeneration of the proximal convoluted tubules was also seen at 8180 mg/m^3. There were no histopathological abnormalities in rats and mice exposed to 414 mg/m^3, and this concentration was considered a clear no-observed-adverse-effect level (Dodd et al., 1982). In a subsequent study, four groups of 14 male and 14 female F-344 rats and 14 male and 14 female B6C3F$_1$ mice were exposed to 0, 205, 1025, or 4100 mg MIBK/m^3 (0, 50, 250, or 1000 ppm, respectively) for 6 h/day (5 days/week for 90 days). No growth retardation or clinical effects were observed in either rats or mice. Clinical observations were made in addition to measurements of body and organ weights (heart, kidneys, liver, lungs, and testes). Haematology, ophthalmology, gross pathology, and histology were also investigated and, in rats, water consumption and urine and serum chemistry analyses were made. Male rats and mice exposed to 4100 mg/m^3 (1000 ppm) showed a slight increase in liver weight (approximately 11%) and in liver weight per body weight ratio, and liver weight was also slightly increased in male mice exposed to 1025 mg/m^3 (250 ppm). However, neither gross nor microscopic hepatic lesions were observed, and urinalysis and serum chemistry values were normal. In male rats exposed to 1025-4100 mg/m^3 (250-1000 ppm), there was an increase in the number of hyaline droplets within the proximal tubular cells of the kidney. No other gross or microscopic renal changes were observed (Dodd & Eisler, 1983; Phillips et al., 1987). It was considered that the hyaline droplet effects produced by MIBK may be specific to the male rat due to the presence of α-2-μglobulin (Phillips et al., 1987).

In studies by Brondeau et al. (1989), groups of five hens were continuously exposed for 50 days by inhalation to either 4100 mg MIBK/m^3 (1000 ppm) or 3520 mg n-hexane/m^3 (1000 ppm) or for 30 days to a mixture of 4100 mg MIBK/m^3 and 3520 mg n-hexane/m^3. Inhalation of n-hexane alone had no effect on hepatic microsomal enzymes, but inhalation of MIBK or the MIBK/n-hexane mixture increased significantly the aniline hydroxylase activity and cytochrome P-450 content of the liver (Abou-Donia et al., 1985b). Inhalation of 2440 to 12 400 mg MIBK/m^3 (595 to 3020 ppm) in rats produced enhancement of the liver cytochrome P-450 content and glutathione-S-transferase activity and also enhanced the ability of 1,2-dichlorobenzene to increase serum glutamate dehydrogenase activity.

Experiments in open-chest cats demonstrated that significant pulmonary hypertension and vasoconstriction, with reduced pulmonary arterial flow, occurred as a result of MIBK inhalation for 5 min at all concentrations tested (410 to 41 000 mg/m^3 (100 to 10 000 ppm)). Systemic arterial pressure and vascular resistance were not significantly affected (Zakhari et al., 1977). Broncho-constriction was also produced by MIBK inhalation for 5 min, the effect being statistically significant at concentrations at or above 2050 mg/m^3 (500 ppm) or 4100 mg/m^3 (1000 ppm) for pulmonary resistance or transpulmonary pressure, respectively. Adult mongrel dogs with open-chest surgery showed pulmonary hypertension at an inhalation concentration of 20.5 mg MIBK/m^3 (5 ppm) for 5 min. At 41 mg/m^3 (10 ppm), myocardial contractility occurred (Zakhari et al., 1977).

8.2.2 Oral

Three daily doses of 375 or 1500 mg MIBK/kg body weight given by gavage to rats reduced the bile flow produced by an intravenous injection of taurocholate (20 mg per kg body weight) (Plaa & Ayotte, 1985). The effect of MIBK on the cholestatic activity of manganese, with or without bilirubin, has also been investigated in male Sprague-Dawley rats. MIBK was administered by gavage in corn oil at doses ranging from 188 to 1502 mg/kg body weight once a day for 1, 3, or 7 days (Vezina et al., 1985; Vezina and Plaa, 1987). MIBK was not cholestatic but, at doses of 375 mg/kg or more, was found to potentiate the cholestasis induced by a manganese-bilirubin combination when this was given 18 h after the 1-day treatment with MIBK. When given for 3 or 7 days, MIBK produced a dose-related enhancement of the cholestasis induced by a manganese-bilirubin combination. A 3-day treatment with MIBK (750 mg/kg) was also shown to potentiate the cholestasis induced by manganese alone. Two known metabolites of MIBK (HMP and 4-methyl-2-pentanol) were also able to potentiate the cholestatic effect of the manganese-bilirubin combination or of manganese alone in male Sprague-Dawley rats. When the metabolite was given by gavage 18 h prior to the administration of the manganese-bilirubin combination, cholestasis was potentiated by 375 mg/kg or 1502 mg/kg (expressed as equivalents of MIBK) of 4-methyl-2-pentanol or HMP, respectively. Lower doses did not decrease bile flow rate. 4-Methyl-2-pentanol was also more effective than HMP as a potentiator following daily treatment for 3 days prior to manganese-bilirubin administration. However, with manganese alone, HMP was more effective. It was suggested that the potentiation of cholestasis may be associated with metabolic induction or might reflect a separate mechanism of action involving a membrane interaction (Vezina & Plaa, 1988).

The toxic effects of MIBK in Sprague-Dawley rats (groups of 30 animals of each sex) were examined following

13 weeks of oral gavage administration at levels of 0, 59, 250, or 1000 mg/kg daily. Body weight, food consumption, organ weight, morbidity, clinical chemistry, haematology, and histopathology evaluations were performed. All surviving animals were killed after 90 days (13 weeks) and 10 animals of each sex per group were examined. Nephrotoxicity was seen as a general nephropathy for both male and female rats administered 1000 mg/kg per day. Although increased liver and kidney weights were observed for males and females at 1000 mg/kg per day, there were no corresponding histopathological lesions present in the liver. The effects seen at 1000 mg/kg per day were present to a significantly lesser extent in the females and males fed 250 mg/kg per day. No effects were observed at 50 mg/kg per day, identifying a no-observed-effect level (Microbiological Associates, 1986).

8.2.3 Parenteral

In studies by Krasavage et al. (1982), rats (strain and number not specified) were given intraperitoneal injections of MIBK or a mixture of methyl ethyl ketone (MEK) and MIBK (9:1 by volume), 5 times/week, for 35 weeks. The dose levels for the first 2 weeks were 10, 30, and 100 mg per kg body weight, and these were then doubled for the remainder of the treatment period. Body weight gain suppression was seen after 3-4 weeks of treatment. The only other effect noted was transient narcosis during the first 4 weeks at the highest dose. Pulmonary vascular effects were observed following an intraperitoneal administration of MIBK to cats (threshold dose 8 mg/kg), but bronchoconstriction was not seen following an intraperitoneal administration of 4-32 mg/kg (Zakhari et al., 1977).

8.2.4 Skin application

In rats exposed dermally to 300-600 mg MIBK/kg per day for 4 months, dose- and time-dependent morphological changes were observed in the skin, brain, liver, adrenals, spleen, and testis. Body temperature decreased and oxygen consumption increased (Malyscheva, 1988).

8.3 Skin, eye, and respiratory irritation; sensitization

8.3.1 Skin irritation

A single 10-h occluded application of MIBK to the shaved skin of rabbits produced erythema, which occurred immediately after the application and persisted for up to 24 h. Daily applications of 10 ml on 10 cm^2 skin for 7 days caused drying and flaking of the surface (Krasavage et al., 1982).

8.3.2 Eye irritation

Undiluted MIBK (0.1 ml) produced some irritation within 10 min when instilled in the rabbit eye. Inflammation and conjunctival swelling occurred within 8 h; the inflammation, swelling, and exudate present at 24 h had disappeared by 60 h (Krasavage et al., 1982).

8.3.3 Respiratory irritation

De Ceaurriz et al. (1981) measured the reflex decrease in respiratory rate in male Swiss OF$_1$ mice as an index of sensory irritation. MIBK caused a concentration-dependent decrease in respiratory rate during a 5-min exposure, and a 50% decrease in respiratory rate (RD$_{50}$) was seen at 13 100 mg/m^3 (3195 ppm). Specht et al. (1940) attributed the decreased respiratory rate to a narcotic effect. It should be recognized that the reduction may not be due to sensory irritation.

8.3.4 Skin sensitization

There are no reports of skin sensitization studies.

8.4 Long-term exposure

No long-term toxicity studies have been reported.

8.5 Reproduction, embryotoxicity, and teratogenicity

In studies by Tyl (1984) and Tyl et al. (1987), groups of 35 pregnant Fischer-344 rats and 30 pregnant CD-1 mice were exposed to 1230, 4100, or 12 300 mg MIBK/m^3 (300,

1000, or 3000 ppm) on days 6-15 (inclusive) of gestation. The animals were sacrificed on days 21 (rats) or 18 (mice) and fetuses examined for external, visceral, and skeletal alterations. In rats, exposure to 12 300 mg/m^3 resulted in maternal toxicity with decreased body weight gain, increased relative kidney weight, decreased food consumption, and fetotoxicity (reduced fetal body weight per litter and delays in skeletal ossification). Clinical signs in dams included loss of coordination, negative toe pinch, paresis, muscular weakness, piloerection, lacrimation, and perioral encrustation. No increase in fetal malformation was observed in any group. At 1230 and 4100 mg/m^3, there was no maternal, embryo, or fetal toxicity, or malformations. However, reduced fetal body weight and delay in some ossification parameters were observed at the lowest dose but not at the intermediate dose level. These effects were attributed to the larger litter size (average 10.8) in this group compared to that in controls (9.5).

In mice, exposure to 12 300 mg/m^3 produced maternal toxicity with increased mortality (3/25), increased absolute and relative liver weights, and fetotoxicity (increased incidence of dead fetuses, reduced fetal body weight per litter, and delayed or reduced ossification). Clinical signs in dams included irregular gait, paresis, hypoactivity, ataxia, negative toe pinch, and lacrimation. There was no treatment-related increase in embryotoxicity or fetal malformations at any exposure concentration tested. No significant treatment-related maternal, embryo, or fetal toxicity (including malformations) was observed at 1230 or 4100 mg/m^3.

8.6 Mutagenicity and related end-points

A summary of the reported data on MIBK mutagenicity is given in Table 6.

8.6.1 Bacterial assays

A pre-incubation assay with Salmonella typhimurium (strains TA98, TA100, TA1537, TA1538) was conducted at dose levels of 0.04 to 4 µg/plate both in the presence and absence of a metabolic activation system prepared from Aroclor-induced rat liver homogenate (S9 fraction).

Table 6. Mutagenicity and related end-points

System	Dose	Response	Reference
Bacterial Assays			
Salmonella typhimurium[a] strains TA98, TA100, TA1537, TA1538	0.04-4 µg/plate	negative	Chemical Manufacturers Association (1984); O'Donoghue et al. (1988)
Salmonella typhimurium[a] strains TA1535, TA1537 TA1538, TA98, TA100	Up to 8000 µg/ml	negative	Brooks et al. (1988)
Eschericia coli strains WP_2 and WP_2 uvr A	Up to 8000 µg/ml	negative	Brooks et al. (1988)
Yeast Assays			
Saccharomyces cerevisiae JDI mitotic gene conversion assay ± rat liver S9	Up to 5 mg/ml	negative	Brooks et al. (1988)
Mammalian cell assays in vitro:			
L51784 TK +/- mouse lymphoma mutation assay (± rat liver S9)	0.001-100 µl/ml (preliminary assay) 0.4-6 µl/ml	negative negative	Chemical Manufacturers Association (1984); O'Donoghue et al. (1988)
Primary rat hepatocytes; unscheduled DNA synthesis (DNA repair)	0.01-100 µl/ml	negative	Chemical Manufacturers Association (1984); O'Donoghue et al. (1988)
Cultured rat liver cells chromosomal damage assay RL_4 cells	Up to 1000 µl/ml (half the dose for 50% inhibition of cell growth)	negative	Brooks et al. (1988)
Balb/3T3; cell transformation assay ± rat liver S9	2-5 µl/ml (-S9) 1-7 µl/ml (+S9)	inconclusive	Chemical Manufacturers Association (1984); O'Donoghue et al. (1988)

Table 6 (contd).

System	Dose	Response	Reference
Mammalian *in vivo* assay			
Mouse (male and female) micronucleus assay (polychromatic erythrocytes)	0.73 ml/kg ip (maximum tolerated dose level)	negative	Chemical Manufacturers Association (1984); O'Donoghue et al. (1988).

a   Preincubation mutation assay incorporating Arochlor-induced rat liver S9

Precautions were taken to prevent the escape of MIBK vapour and assure prolonged exposure of the bacteria to the test substance. MIBK did not cause an increase in reverse gene mutation (Chemical Manufacturers Association 1984; O'Donoghue et al., 1988).

Both MIBK and the oxidative metabolite 4-hydroxy-4-methyl-2-pentanone (HMP) (see section 6.1) were tested for mutagenicity in Salmonella typhimurium strains TA98, TA100, TA1535, TA1537, and TA1538, and in *Eschericia coli* strains WP_2 and WP_2 *uvr* A (Brooks et al., 1988). Aroclor-induced rat liver S9 fraction was included. No induction of reverse gene mutation was observed up to a maximum concentration of 8000 µg MIBK/ml (pre-incubation assay in a sealed container) or 4000 µg HMP/plate (plate incorporation assay).

8.6.2 *Yeast assay for mitotic gene conversions*

MIBK and the metabolite HMP were assayed for mitotic gene conversion using log-phase cultures of the yeast *Saccharomyces cerevisiae* JD1 (Brooks et al., 1988). Compounds were tested up to a concentration of 5 mg/ml in the presence and absence of rat liver S9 fraction in a sealed container for 18 h. Neither compound induced mitotic gene conversion.

8.6.3 L5178Y TK+/- mouse lymphoma assay

A preliminary assay was carried out in the presence and absence of a metabolic activation system at doses of 0.001-100 µl/ml. The non-activated cultures showed 3 to 157% total relative growth, while the cultures containing the rat liver S9 fraction had a relative growth of 23-95% compared with untreated control cultures. No increase in mutation frequencies was observed in cultures containing the metabolic activation system, but in the non-activated cultures, a 2-fold increase above controls was seen at two non-consecutive doses. An increase in mutation frequency of approximately 5 times the concurrent control occurred at one test concentration, but this concentration also caused 97% cell death. In the absence of a dose-related effect, this result was considered equivocal. A repeat assay was performed using duplicate cultures and a narrower range of doses (0.4-6 µl/ml). The total relative growth ranged from 1 to 80% in non-metabolically activated cells and from 28 to 63% in cultures that contained the S9 fraction. None of the activated cultures revealed increased mutation frequencies. A borderline positive result was found at 6 µl/ml, but different mutation frequencies occurred in the duplicate cultures and 96-99% of the cells were killed (Chemical Manufacturers Association, 1984; O'Donoghue et al., 1988).

8.6.4 Unscheduled DNA synthesis in primary rat hepatocytes in vitro

When MIBK was tested at five dose levels ranging from 0.01 µl/ml to 100 µl/ml in a single assay, there was an increase of less than 5 fold in labelled nuclear grains in cells treated with MIBK compared with cells of the solvent control plates (Chemical Manufacturers Association, 1984; O'Donoghue et al., 1988). Since the value did not exceed that of the negative control by two standard deviations of the control value, it was considered that, under the conditions tested, MIBK did not cause a significant increase in the nuclear grain count.

8.6.5 Mouse micronucleus assay

Male and female mice were administered MIBK by ip injection at the maximum tolerated dose level of 0.73 ml/kg body weight, and bone marrow polychromatic erythrocytes were estimated 12, 24, and 48 h later. There were no significant differences between the treated and control animals in the ratio of polychromatic to normochromatic erythrocytes. The number of micronucleated polychromatic erythrocytes per 1000 cells was not significantly increased in the MIBK-treated animals (Chemical Manufacturers Association, 1984; O'Donoghue et al., 1988).

8.6.6 Assay for structural chromosome damage

MIBK (purity > 98.5%) and the metabolite HMP were tested (24-h exposure) in cultured rat liver cells (RL_4) for the ability to induce chromosomal damage. Metabolic activation with S9 mix was not used because RL_4 cells are metabolically competent. The concentrations of MIBK employed were 0.125, 0.25, and 0.5 times the concentration required for 50% inhibition of cell growth. Maximum concentrations tested were thus 1000 µg MIBK/ml and 4000 µg HMP/ml (this HMP level was equivalent to the concentration required for >60% growth inhibition). Incubations with MIBK were sealed to prevent loss by evaporation. MIBK did not produce chromosomal damage, but HMP gave a small increase (which was not dose related) in chromatid damage within the concentration range 2000-4000 µg/ml. It should be noted that HMP did not induce reverse gene mutation in bacteria or mitotic gene conversion in yeast (Brooks et al., 1988).

8.6.7 Cell transformation assay

In studies reported by the Chemical Manufacturers Association (1984) and O'Donoghue et al. (1988), MIBK was tested in the Balb/3T3 (clone A31-1) morphological transformation assay. Doses of 2.4, 3.6, and 4.8 µl MIBK/ml were added to the culture medium in the absence of a metabolic activation system, and 1, 2, and 4 µl MIBK/ml were added in the presence of such a system (Aroclor-induced rat liver S9 fraction). MIBK produced a positive response in the non-activated cultures only (4.8 µl MIBK per ml gave 3 type III foci in 15 dishes). A confirmatory study was conducted with doses of 2, 3, 4, and 5 µl/ml and 4, 5, 6, and 7 µl/ml, respectively, in the presence and absence of S9 fraction. No significant increase in the number of transformed foci was found in this study, either in the presence or absence of the metabolic activation system. Thus, the effect of MIBK on cell transformation was not reproducible in the two assays and the ambiguity of the results makes them unreliable.

8.7 Carcinogenicity

No carcinogenicity studies have been reported.

8.8 Neurotoxicity

In a study by Krasavage et al. (1982), rats were given ip injections of MIBK, or a mixture of methyl ethyl ketone and MIBK (9:1 by volume), 5 times/week, for 35 weeks. The dose levels of 10, 30, and 100 mg/kg body weight were doubled after 2 weeks of treatment. Transient anaesthesia was noted during the first 4 weeks in the highest dose group, but there was no evidence of peripheral neuropathy. In dogs administered 300 mg MIBK/kg body weight per day subcutaneously (sc) for 11 months, electromyographic examination showed no evidence of neurotoxicity.

Cats treated subcutaneously with 150 mg MIBK/kg body weight per day or a mixture of methyl ethyl ketone/MIBK (9:1) twice daily, 5 times/week, for up to 8.5 months showed no evidence of nervous system damage (Spencer & Schaumburg, 1976). In beagle dogs receiving similar treatment, there were no neurotoxic changes (Krasavage et al., 1982).

Groups of male rats were exposed to 5330 mg/m^3 (1300 ppm) methyl n-butyl ketone for 4 months or 6150 mg/m^3 (1500 ppm) MIBK for 5 months (Spencer et al., 1975). Methyl n-butyl ketone produced a toxic distal axonopathy. MIBK produced minimal distal axonal changes, but 3% methyl n-butyl ketone was present as a contaminant in the MIBK, and the design of the cages used may have caused compression neuropathy (Spencer et al., 1975; Spencer & Schaumburg, 1976). Animals exposed to MIBK showed slight signs of narcosis, but body weight gain was normal, and, at 5 months, there were no clinical signs of neurological dysfunction. Rats exposed for 3 h to 102 mg MIBK/m^3 (25 ppm) showed a 58% increase in pressor lever response, which had not returned to control levels 7 days after the end of exposure (Geller et al., 1978). The maximum motor-fibre conduction velocity in the tail nerve decreased markedly when male rats were treated with methyl *n*-butyl ketone (401 mg/kg, 5 times/week for 55 weeks) but not when they were treated with MIBK (601 mg/kg, 5 times/week for 55 weeks). However, treatment with MIBK (201 mg/kg) facilitated the neurotoxic effect of methyl *n*-butyl ketone (401 mg/kg) possibly due to the demonstrated ability of MIBK to increase the metabolic activity of 10 000 g liver supernatants towards both MIBK and methyl *n*-butyl ketone (Nagano et al., 1988).

Discriminatory behaviour and memory in baboons was not affected by exposures of 82-164 mg/m^3 (20-40 ppm) (Geller et al., 1978). Geller et al. (1979) reported an effect on accuracy of performance of tasks in a "delayed match-to-sample discrimination-test" in baboons exposed for 7 days to 205 mg/m^3 (50 ppm) MIBK, but there was no change in response when MIBK was combined with methyl ethyl ketone at 295 mg/m^3 (100 ppm).

De Ceaurriz et al. (1984) exposed male Swiss OF1 mice to an atmosphere containing MIBK and measured the total duration of immobility during a 3-min "behavioural despair" swimming test. At concentrations of 2714, 3104, 3309, and 3657 mg/m^3 (662, 757, 807, and 892 ppm), a dose-dependent decrease of mobility was observed (25, 38, 46, and 70% respectively). The authors noted that the mean active level of MIBK was lower for this neurobehavioural effect (3292 mg/m^3 (893 ppm) for 50% inhibition of immobility) than for an effect on sensory irritation (13 100 mg/m^3 (3195 ppm) for 50% inhibition of respiratory rate) (De Ceaurriz et al., 1981, section 8.3.3).

In inhalation studies on hens (five per group) into the effect of MIBK on *n*-hexane-induced neurotoxicity, a continuous exposure period of 90 days was followed by a 30-day observation period (Abou-Donia et al., 1985b). One group of hens was exposed to 3520 mg *n*-hexane/m^3 (1000 ppm) and another group to 4100 mg MIBK/m^3 (1000 ppm). Four additional groups were exposed simultaneously to 3520 mg *n*-hexane/m^3 (1000 ppm) and 410, 1025, 2050, or 4100 mg MIBK/m^3 (100, 250, 500, or 1000 ppm, respectively). A control group was exposed to ambient air in an exposure chamber. Hens continuously exposed to 4100 mg MIBK/m^3

developed weakness of the legs with subsequent recovery. Inhalation of 3520 mg n-hexane/m^3 produced mild ataxia. Exposure to 3520 mg n-hexane/m^3 together with 1025, 2050, or 4100 mg MIBK/m^3 resulted in signs of neurotoxicity including paralysis, the severity of which depended on the MIBK concentration. Hens continuously exposed to the 3520/410 mg/m^3 n-hexane/MIBK mixture exhibited severe ataxia throughout the observation period. Histopathological examination of hens exposed to the n-hexane/MIBK mixture showed large swollen axons and degeneration of the axon and myelin of the spinal cord and peripheral nerves. There were no histopathological abnormalities in the central nervous system of hens exposed only to MIBK. This demonstrates that MIBK potentiates the neurotoxic action of n-hexane. In a subsequent study, in which hens were exposed for 50 days to MIBK at 4100 mg/m^3 (1000 ppm) or to n-hexane, it was suggested that the potentiation by MIBK of n-hexane neurotoxicity is related to the induction by MIBK of liver microsomal cytochrome P-450, resulting in increased metabolism of n-hexane to its neurotoxic metabolites (sections 6.1, 8.2.1). The findings also suggest that the neurotoxicity of technical methyl butylketone (methyl n-butyl ketone/MIBK (7/3)) was correctly attributed to the methyl n-butyl ketone component (Abdo et al., 1982). Simultaneous treatment of hens with 41, 205, or 410 mg/m^3 (10, 50, or 100 ppm) technical methyl butyl ketone, 5 days/week for 90 days, with a dermal application of technical grade 0-ethyl-0-4-nitrophenyl phenylphosphonothioate (EPN, 1.0 mg/kg, 85%) greatly enhanced the neurotoxic effects. It was proposed that MIBK partially contributed to this potentiation by inducing cytochrome P-450 and thus enhancing the formation of neurotoxic products from methyl-n-butyl ketone and EPN (Abou-Donia et al., 1985a).

The effect of MIBK on the duration of ethanol-induced loss of righting reflex and on ethanol elimination has been studied in mice. MIBK was dissolved in corn oil and injected ip 30 min before ethanol (4 g/kg ip). At a dose of 501 mg/kg, MIBK significantly prolonged the duration of ethanol-induced loss of righting reflex. The concentrations of ethanol in blood and brain on return of the righting reflex were similar in MIBK-treated and control animals (Cunningham et al., 1989).

8.9 In vitro toxicity assays

In contrast to methyl n-butyl ketone and n-hexane, MIBK caused little or no cytopathological or growth-inhibiting effects in cultured mouse neuroblastoma cells (Selkoe et al., 1978). MIBK in N,N-dimethylacetamide reduced the activity of mouse liver alcohol dehydrogenase *in vitro* (Cunningham et al., 1989). MIBK was found to inhibit the sulfhydryl-dependent creatine kinase and adenylate kinase enzymes *in vitro* but not to the same extent as did the neurotoxic agent, methyl-n-butyl ketone (Lapin et al., 1982).

9. EFFECTS ON MAN

9.1 Acute toxicity

In a study on the sensory threshold, Silverman et al. (1946) exposed 12 volunteers of both sexes to various concentrations of MIBK for a 15-min period. This period permitted an accurate observation of olfactory fatigue and increasing or decreasing irritation of mucous membranes. The sensory response limit was 410 mg/m^3 (100 ppm). The majority of the subjects found the odour objectionable at 820 mg/m^3 (200 ppm), and the vapour irritated the eyes. The low odour threshold (1.64 mg/m^3) (Ruth, 1986) and the irritant effects can provide warning of high concentrations. Because of its low viscosity, MIBK may, when swallowed, also be aspirated into the lungs causing a chemical pneumonitis (Panson & Winek, 1980).

9.2 Short-term exposure

Workers exposed to 410 mg MIBK/m^3 (100 ppm) complained either of headache and nausea or of respiratory irritation (Elkins, 1959). Tolerance was said to be acquired during the working week but was lost over the weekend. Reduction of the exposure to 82 mg/m^3 (20 ppm) largely eliminated the complaints. In the study of Hjelm et al. (1990) (see section 6.3) on human volunteers, CNS symptoms (headache and/or vertigo and/or nausea) were reported at 2-h exposure levels of 10-200 mg MIBK/m^3 (2.4-48.8 ppm). There were no significant effects from exposure on the performance of a reaction time task or a test of mental arithmetic.

9.3 Eye and respiratory irritation

Exposure to a concentration of 820 mg MIBK/m^3 (200 ppm) for a 15-min period caused eye irritation in 12 human volunteers (Silverman et al., 1946). Undiluted MIBK splashed in the eyes may cause painful irritation (Shell, 1957). A group of workers exposed to 410 mg MIBK/m^3 (100 ppm) complained of respiratory tract irritation, but there were no complaints at 82 mg/m^3 (20 ppm) (Elkins, 1959).

In the study of Hjelm et al. (1990) (see section 6.3) on human volunteers, irritation particularly of the nose and throat was reported at 2-h exposure levels of 10, 100, and 200 mg/m^3 (2.4, 24.4, and 48.8 ppm).

9.4 Long-term exposure

In workers exposed to up to 2050 mg MIBK/m^3 (500 ppm) for 20-30 min per day and to 328 mg/m^3 (80 ppm) for much of the remainder of the working day, over half of the 19 workers complained of weakness, loss of appetite, headache, eye irritation, stomach ache, nausea, vomiting, and sore throat. A few of the workers experienced insomnia, somnolence, heartburn, intestinal pain, and some unsteadiness. Four workers had slightly enlarged livers and six had a nonspecific colitis. Clinical chemistry examination revealed no abnormalities in any of the workers. Five years later, work practices had greatly improved, the highest MIBK concentration was 410-430 mg/m^3 (100-105 ppm), and the general concentration was 205 mg/m^3 (50 ppm). A few workers still complained of gastrointestinal and central nervous system effects, and slight liver enlargement had persisted in two workers, but other symptoms had disappeared (Armeli et al., 1968).

9.5 Placental transfer

MIBK was detected in maternal and umbilical cord blood samples from 11 patients (Dowty et al., 1976).

9.6 Neurotoxicity

A few isolated cases of peripheral neuropathy have been reported after exposure to spray paint or lacquer thinner that apparently contained MIBK and other hydrocarbon solvents, including neurotoxic agents (Oh & Kim, 1976; Aubuchon et al., 1979).

10. EVALUATION OF HUMAN HEALTH RISKS AND EFFECTS ON THE ENVIRONMENT

10.1 Evaluation of effects on the environment

MIBK is not likely to persist in the environment. It will slowly volatilize from soil and water and is readily biodegraded in fresh and salt water. In the atmosphere, MIBK is estimated to be degraded by OH· radicals with a half-life of approximately 14 h. MIBK is not expected to bioaccumulate and has a low toxicity for microorganisms, fish, algae, and aquatic invertebrates. Only in cases of accidental spillage or inappropriate disposal of wastes into the environment are levels of MIBK likely to cause toxicity to organisms in the environment.

10.2 Evaluation of health risks for humans

The general population is exposed to low levels of MIBK. Only small quantities have been detected in food, drinking-water, and other beverages (baked goods, 10.9 mg/kg; frozen dairy products, 11.5 mg/kg; gelatins, puddings, 10.9 mg/kg; beverages, 10.2 mg/kg). For general population exposure, maximum ambient air concentrations in the range of 0.1 to 0.2 mg/m^3 have been defined by two countries.

Occupational exposure occurs particularly in the production and use of lacquers, paints, and extraction solvents. The major route of entry is by inhalation. The low odour threshold (1.64 mg/m^3) and the irritant effects can provide warning of high concentrations. Exposure to levels of 10-410 mg/m^3 (2.4-100 ppm) produced perceptible irritation of either the eyes, nose, or throat, and 820 mg/m^3 (200 ppm) produced discomfort. Symptoms such as headache, nausea, and vertigo also occurred at a level of 10-410 mg/m^3 (2.4-100 ppm). There were no significant effects from a 2-h exposure of up to 200 mg/m^3 (50 ppm) on a simple reaction time task or test of mental arithmetic.

In the single report concerning long-term occupational exposure, where workers were exposed to 2050 mg MIBK/m^3

(500 ppm) for 20-30 min per day and to 328 mg/m^3 (80 ppm) for much of the remainder of the working day, more than half of the 19 workers complained of weakness, loss of appetite, headache, eye irritation, stomach ache, nausea, vomitting, and sore throat. A few workers experienced insomnia, somnolence, and some unsteadiness. Four had slightly enlarged livers and six had nonspecific colitis. Five years later, work practices had greatly improved and the highest concentrations were reduced to about one fifth of the previous level. A few workers still complained of irritation of the eyes and upper respiratory tract as well as gastrointestinal and central nervous system symptoms. Prolonged skin contact with MIBK caused irritation and flaking of the skin.

In animal studies, acute systemic MIBK toxicity is low by the oral and inhalation routes. In a 90-day study, Sprague-Dawley rats were given MIBK by gavage at doses of 50, 250, or 1000 mg/kg body weight per day. Lethargy was noted in the highest-dose group and males showed reduced body weight gain. In this group there was generalized nephropathy, with an increase in relative kidney weight and hepatomegaly. Relative kidney weight was also increased in the animals fed 250 mg/kg per day, and slight hepatomegaly was reported in male rats only. There were no histopathological lesions in the liver or other tissues at any dose level. It was concluded that the NOEL was 50 mg/kg per day. In 90-day inhalation studies on rats and mice, concentrations of up to 4100 mg/m^3 (1000 ppm) did not result in any life-threatening signs of toxicity. However, compound-related reversible morphological changes in the liver and kidney were reported. Levels of 4100 mg/m^3 produced evidence of central nervous system depression. MIBK was capable of increasing liver weight (at > 1025 mg/m^3 (250 ppm)) and inducing hepatic microsomal metabolism. This may be the explanation for the exacerbation of haloalkane toxicity and the potentiation of the neurotoxicity of *n*-hexane. In 90-day studies with mice, rats, dogs, and monkeys, only male rats developed hyaline droplets in the proximal tubules of the kidney (hyaline droplet toxic tubular nephrosis). This effect in male rats was reversible and of doubtful significance for humans. MIBK reduces the activity of mouse liver alcohol dehydrogenase *in vitro*. It has also been found to potentiate the cholestatic effects of manganese given with or without bilirubin.

In rats and mice exposed to MIBK by inhalation at concentrations of 1230, 4100, or 12 300 mg/m^3 (300, 1000, or 3000 ppm) on days 6 to 15 of gestation and sacrificed on day 21 (rats) or day 18 (mice), marked maternal toxicity was observed at the highest concentration in both species. This concentration produced fetotoxicity (reduced fetal body weight and delayed ossification) but was not embryotoxic or teratogenic. At 4100 and 1230 mg/m^3 there was no maternal toxicity and no evidence of embryotoxicity, fetotoxicity, or teratogenicity.

MIBK did not induce gene mutation in bacterial test systems (Salmonella typhimurium and *Escherichia coli*) either with or without metabolic activation. Negative results were also obtained in tests (both with and without metabolic activation) for mitotic gene conversion in yeast (*Saccharomyces cerevisiae*) and in gene mutation tests using cultured mammalian cells (mouse lymphoma). *In vitro* assays for unscheduled DNA synthesis in primary rat hepatocytes and for structural chromosome damage in cultured rat liver cells (RL4) were negative. An *in vivo* micronucleus test in mice was negative. These data indicate that MIBK is not genotoxic.

11. RECOMMENDATIONS

At the levels of MIBK to which the general human population is exposed, there is unlikely to be any hazard. In the occupational health context, where the major route of exposure is by inhalation, atmospheric levels should be kept below the recommended occupational exposure limits by suitably designed work processes and engineering controls, including ventilation. Skin and eye contamination should be avoided. Suitable protective clothing and respiratory protection should be readily available for use in enclosed spaces, in emergencies, and for certain maintenance operations. MIBK is inflammable and should be handled accordingly.

MIBK has low toxicity for microorganisms and fish, and its half-life in the environment is short. Consequently, there is no risk to the environment provided there are adequate controls to minimize emissions. Large-scale release could have local adverse effects on the environment.

12. FURTHER RESEARCH

1. MIBK affects a number of enzyme systems. Therefore, it can significantly influence the biotransformation of xenobiotics that are metabolized by these enzymes. Since humans are usually exposed to more than one compound, studies on the combined effects of mixtures containing MIBK should be undertaken.

2. There is very little information available on the dose-response relationships for the effects of MIBK on the human central nervous system (e.g., reaction time, behavioural effects), on the upper airways and mucous membranes, or on kidney function. More information on toxicokinetics is needed for MIBK alone and in mixture with other solvents. The skin penetration of MIBK should be assessed.

3. Epidemiological studies should be undertaken to elucidate the effects on the nervous system of long-term exposure to moderate concentrations of MIBK alone or in mixture with other solvents.

REFERENCES

ABDO, K.M., GRAHAM, D.G., TIMMINS, P.R., & ABOU-DONIA, M.B. (1982) Neurotoxicity of continuous (90 days) inhalation of technical grade methyl butyl ketone in hens. J. Toxicol. environ. Health, 9: 199-215.

ABOU-DONIA, M.B., LAPADULA, D.M., CAMPBELL, G., & ABDO, K.M. (1985a) The joint neurotoxic action of inhaled methyl butyl ketone vapour and dermally applied 0-ethyl 0-4-nitrophenyl phenylphosphonothioate in hens: potentiating effect. Toxicol. appl. Pharmacol., 79: 69-82.

ABOU-DONIA, M.B., LAPADULA, D.M., CAMPBELL, G., & TIMMONS, P.R. (1985b) The synergism of n-hexane-induced neurotoxicity by methyl isobutyl ketone following subchronic (90 days) inhalation in hens: induction of hepatic microsomal cytochrome P-450. Toxicol. appl. Pharmacol., 81: 1-16.

ANALYTICAL QUALITY CONTROL (1972) Handbook for analytical quality control in water and waste water laboratories, Cincinnati, Ohio, National Environmental Research Centre.

ARMELI, G., LINARI, F., & MARTORANO, G. (1968) [Clinical and haematochemical examinations in workers exposed to the action of a higher ketone (MIBK) repeated after 5 years.] Lav. Um., 20: 418-424 (in Italian).

ATKINSON, R., ASCHMANN, S.M., CARTER, W.P.L., & PITTS, J.N., Jr (1982) Rate constants for the gas-phase reaction of OH radicals with a series of ketones at 299 $\pm$ 2 ° K. Int. J. chem. Kinet., 14: 839-847.

AUBUCHON, J., ROBINS, H.I., & VISESKUL, C. (1979) Peripheral neuropathy after exposure to methyl isobutyl ketone in spray paint. Lancet, August 18: 363-364.

AUSTERN, B.M., DOBBS, R.A., & COHEN, J.M. (1975) Gas-chromatographic determination of selected organic compounds added to wastewater. Environ. Sci. Technol., 9: 588-590.

BAMBERGER, R.L., ESPOSITO, G.G., JACOBS, B.W., PODOLAK, G.E., & MAZUR, J.F. (1978) A new personal sampler for organic vapors. Am. Ind. Hyg. Assoc. J., 39: 701-708.

BASU, P., CARPENTER, J., CHEN, C., NELSON, H., PERRY, W., SHOCLET, A., TAYLOR, D., & ZAFRAN, F. (1968) Human exposure assessment: Methyl isobutyl ketone, Washington, DC, US Environmental Protection Agency (EPA contract 68-01.4839).

BATYROVA, T.F. (1973) Substantiation of the maximum permissible concentration of methylisobutyl ketone in air or workrooms. Gig. Tr. prof. Zabol., 17(11): 52- 53.

BELLANCA, J.A., DAVIS, P.L., DONNELLY, B., DAL CORTIVO, L.A., & WEINBERG, S.B. (1982) Detection and quantitation of multiple volatile compounds in tissues by GC and GC/MS. J. anal. Toxicol., 6: 238-240.

BRANCHFLOWER, R.V., SCHULICK, R.D., GEORGE, J.W., & POHL, L.R. (1983) Comparison of the effects of methyl n-butyl ketone and phenobarbital on rat liver cytochromes P-450 and the metabolism of chloroform to phosgene. Toxicol. appl. Pharmacol., 71: 414-421.

BRIDIE, A.L., WOLFF, C.J.M., & WINTER M. (1979a) The acute toxicity of some petrochemicals to goldfish. Water Res., 13: 623-626.

BRIDIE, A.L., WOLFF, C.J.M., & WINTER, M. (1979b) BOD and COD of some petrochemicals. Water Res., 13: 627-630.

BRINGMANN, G. & KUHN, R. (1977a) [Findings concerning the harmful effect of water-endangering substances on *Daphnia magna*.] Z. Wasser Abwasser Forsch., 10: 161-166 (in German).

BRINGMANN, G. & KUHN, R. (1977b) [Limit values for the harmful effect of water-endangering substances on bacteria (*Pseudomonas putida*) and green algae (*Scenedesmus quadricauda*).] Z. Wasser Abwasser Forsch., 10: 87-98 (in German).

BRINGMANN, G. & KUHN, R. (1978) [Limit values for the harmful effect of water-endangering substances on blue-green algae (*Microcystis aeruginosa*) and green algae (*Scenedesmus quadricauda*) in the cell multiplication inhibitor test.] Vom Wasser, 50: 45-60 (in German).

BRINGMANN, G. & KUHN, R. (1981) [Comparison of the effects of pollutants on flagellates and ciliates, and/or on holozoic bacteria-eating and saprozoic protozoa.] GWF-Wasser Abwasser, 122: 308-313 (in German).

BRINGMANN, G. & KUHN, R. (1982) [Results of toxic action of water pollutants on *Daphnia magna straus* tested by an approved standardized procedure.] Z. Wasser Abwasser Forsch., 15: 1-6 (in German).

BRONDEAU, M.T., BAN, M., BONNET, P., GUENIER, J.P., & DE CEAURRIZ, J. (1989) Acetone compared to other ketones in modifying the hepatotoxicity of inhaled 1,2-dichlorobenzene in rats and mice. Toxicol. Lett. 49: 69-78.

BROOKS, T.M., MEYER, A.L., & HUTSON, D.H. (1988) The genetic toxicology of some hydrocarbon and oxygenated solvents. Mutagenesis, 3: 227-232.

BROWN, K.W. & DONNELLY, K.C. (1988) An estimation of the risk associated with the organic constituents of hazardous and municipal waste landfill leachates. Hazardous Waste Hazardous Mater., 5(1): 1-30.

BROWN, R.H. & PURNELL, C.J. (1979) Collection and analysis of trace organic vapour pollutants in ambient atmospheres. J. Chromatogr., 178: 79-90.

BURNHAM, A.K., CALDER, G.V., FRITZ, J.S., JUNK, G.A., SVEC, H.J., & WILLIS, R. (1972) Identification and estimation of neutral organic contaminants in potable water. Anal. Chem., 44: 139-142.

CALL, D.J., BROOKE, L.T., KNUTH, M.L., POIRIER, S.H., & HOGLUND, M.D. (1985) Fish subchronic toxicity prediction model for industrial organic chemicals that produce narcosis. Environ. Toxicol. Chem., 4: 335-341.

CEC (1976) Analysis of organic micropollutants in water, Luxembourg, Commission of the European Communities (Cost 64b bis).

CFR (1987) Code of Federal Regulations. Methods of analysis for organic chemicals in groundwater at hazardous waste sites, Washington, DC, US Government Printing Office (Appendix IX, 40 CFR, Part 264).

CHEMICAL MANUFACTURERS ASSOCIATION (1984) Ketones Program Panel, Vol. 1 - Methyl isobutyl ketone: Mutagenicity and teratology studies, Washington, DC, Chemical Manufacturers Association.

CORWIN, J.F. (1969) Volatile oxygen-containing organic compounds in sea water: determination. Bull. mar. Sci., 19: 504-509.

COX, R.A., DERWENT, R.G., & WILLIAMS, M.R. (1980) Atmospheric photooxidation reactions. Rates, reactivity and mechanism for reaction of organic compounds with hydroxyl radicals. Environ. Sci. Technol., 14: 57-61.

CUNNINGHAM, J., SHARKAWI, M., & PLAA, G.L. (1989) Pharmacological and metabolic interactions between ethanol and methyl n-butyl ketone, methyl isobutyl ketone, methyl ethyl ketone, or acetone in mice. Fundam. appl. Toxicol., 13: 102-9.

DE CEAURRIZ, J., MICILLINO, J.C., BONNET, P., & GUENIER, J.P. (1981) Sensory irritation caused by various industrial airborne chemicals. Toxicol. Lett., 9: 137-143.

DE CEAURRIZ, J., MICILLINO, J.C., MARIGNAC, B., BONNET, P., MULLER, J., & GUENIER, J.P. (1984) Quantitative evolution of sensory irritating and neurobehavioural properties of aliphatic ketones in mice. Food Chem. Toxicol., 22, 545-549.

DICK, R., DANKOVIC, D., SETZER, J., PHIPPS, F., & LOWRY, L. (1990) Body burden profiles of methyl ethyl ketone and methyl isobutyl ketone exposure in human subjects. Toxicologist, 10: 122.

DIVINCENZO, G.D. & KRASAVAGE, W.J. (1974) Serum ornithine carbamyl transferase as a liver response test for exposure to organic solvents. Am. Ind. Hyg. Assoc. J., 35: 21-29.

DIVINCENZO, G.D., KAPLAN, C.J., & DEDINAS, J. (1976) Characterization of the metabolites of methyl n-butyl ketone, methyl iso-butyl ketone and methyl ethyl ketone in guinea pig serum and their clearance. Toxicol. appl. Pharmacol., 36: 511-522.

DODD, D.E. & EISLER, D.L. (1983) Methyl isobutyl ketone ninety-day inhalation study on rats and mice, Washington, DC, Chemical Manufacturers Association (Bushy Run Research Center, Report 46-504 submitted to US EPA).

References

DODD, D.E., LONGO, L.C., & EISLER, D.L. (1982) Nine-day vapour inhalation study on rats and mice, Washington, DC, Chemical Manufacturers Association (Bushy Run Research Center, Report 45-501 submitted to US EPA).

DOWTY, B.J., LASETER, J.L., & STORER, J. (1976) The transplacental migration and accumulation in blood of volatile organic constituents. Pediatr. Res.,10: 696-701.

ECDIN (1990) Data bank on environmental chemicals, Ispra (Varese), Establishment, Joint Research Centre of the Commission of the European Communities.

ELKINS, H.B. (1959) Chemistry of industrial toxicology, New York, John Wiley and Sons, p. 121.

ELLISON, W.K. & WALLBANK, T.E. (1974) Solvents in sewage and industrial waste waters. Identification and determination. Water Pollut. Control, 73: 656-672.

ELOFSSON, S.A., GAMBERALE, F., HINDMARSH, T., IREGREN, A., ISAKSSON, A., JOHNSSON, I., KNAVE, B., LYDAHL, E., MINDUS, P., PERSSON, H.E., PHILIPSON, B., STEBY, M., STRUWE, G., SODERMAN, E., WENNBERG, A., & WIDEN, L. (1980) Exposure to organic solvents: a cross-sectional epidemiologic investigation on occupationally exposed car and industrial spray painters with special reference to the nervous system. Scand. J. Work Environ. Health, 6: 239-273.

FAWELL, J.K. & HUNT, S. (1981) Organic micropollutants in drinking water, Medmenham, Water Research Centre (Technical Report No. 159).

FERNANDES, M. (1985) Methodology for the analysis of volatile compounds in food packaging materials. Coletanea Inst. Technol. Aliment., 15: 49-59.

FIRE PREVENTION (1981) Information sheets on hazardous materials, London, London Fire Protection Association, p. 47 (H97 No. 140).

FRANCIS, A.J., IDEN, G.T., NINE, B.J., & CHANG, C.K. (1980) Characterization of organics in leachates from low level radioactive waste disposal sites. Nucl. Technol., 50: 158-163.

FROSTLING, H., HOFF, A., JACOBSSON, S., PFAFFLI, P., VAINIOTALO, S., ZITTING, A., & TECHN, D. (1984) Analytical, occupational and toxicologic aspects of the degradation products of polypropylene plastics. Scand. J. Work Environ. Health., 10: 163-169.

GARMAN, J.R., FREUND, T., & LAWLESS, E.W. (1987) Testing for groundwater contamination at hazardous waste sites. Chromatogr. Sci., 25: 328-344.

GELLER, I., ROWLANDS, J.R., & KAPLAN, H.L. (1978) Effects of ketones on operant behaviour of laboratory animals. In: Voluntary inhalation of industrial solvents, Washington, DC, US Department of Health, Education and Welfare, p. 363 (DHEW Publication No. 79-779).

GELLER I., GAUSE, E., KAPLAN, H., & HARTMANN, R.J. (1979) Effects of acetone, methyl ethyl ketone, and methyl isobutyl ketone on a match-to-sample task in the baboon. Pharmacol. Biochem. Behav., 11: 401-406.

HAMPTON, C.V., PIERSON, W.R., HARVEY, T.M., UPDEGROVE, W.S., & MARANO, R.S. (1982) Hydrocarbon gases emitted from vehicles on the road. 1. A qualitative gas chromatography/mass spectroscopy survey. Environ. Sci. Technol., 16: 287-298.

HANNINEN, H., ESKELINEN, L., HUSMAN, K., & NURMINEN, M. (1976) Behavioural effects of long-term exposure to a mixture of organic solvents. Scand. J. Work Environ. Health, 4: 240-255.

HJELM, E.W., HAGBERG, M., IREGREN, A., & LÖF, A. (1990) Exposure to methyl isobutyl ketone: toxicokinetics and occurrence of irritative and CNS symptoms in man. Int. Arch. occup. environ. Health, 62: 19-26.

IRPTC (1990) IRPTC legal file, Geneva, International Register of Potentially Toxic Chemicals, United Nations Environment Programme.

JUHNKE, I. & LÜDEMANN, D. (1978) [Results of the testing of 200 chemical compounds for acute fish toxicity in the orfe test.] Z. Wasser Abwasser Forsch., 11(5): 161-164 (in German).

KEITH, L.H. (1974) Chemical characterization of industrial waste waters by gas chromatography-mass spectrometry. Sci. total Environ., 3: 87-102.

KRASAVAGE, W.J., O'DONOGHUE, J.L., & DIVINCENZO, G.D. (1982) Methyl isobutyl ketone. In: Clayton, G.D. & Clayton, F.E., ed. Patty's industrial hygiene and toxicology, New York, John Wiley and Sons, Vol. 2e, pp. 4747-4751.

KRISTENSSON, J. & BEVING, H. (1987) A study of painters occupationally exposed to water and solvent based paints, Luxembourg, Commission of the European Communities, pp. 71-72 (EUR. 10555).

LANDE, S.S., DURKIN, P.R., CHRISTOPHER, D.H., HOWARD, P.H., & SAXENA, J. (1976) Investigation of selected potential environmental contamination: ketonic solvents, Syracuse, New York, Center for Chemical Hazard Assessment Research Corporation, p. 252.

LAPIN, E.P., WEISSBARTH, S., MAKER, H.S., & LEHRER, G.M. (1982) The sensitivities of creatine and adenylate kinases to the neurotoxins acrylamide and metyl n-butyl ketone. Environ. Res., 28: 21-31.

LEO, A. & WEININGER, D. (1984) Medicinal chemistry report, Claremont, California, Pomona College.

LEVIN, J.-O. & CARLEBORG, L. (1987) Evaluation of solid sorbents for sampling ketones in work-room air. Ann. occup. Hyg., 31: 31-38.

LIPNICK, R.L., WATSON, K.R., & STRAUSZ, A.K. (1987) A QSAR study of the acute toxicity of some industrial organic chemicals to goldfish. Narcosis, electrophile and proelectrophile mechanisms. Xenobiotica, 17, 1011-1025.

References

MACEWEN, J.D., VERNOT, E.H., & HAUN, C.C. (1971) Effects of 90-day continous exposure to methyl isobutyl ketone on dogs, monkeys, and rats, Ohio, Wright-Patterson AFB, Aerospace Medical Research Laboratory (Report No. AMRL TR-71-65).

MACKAY, D. & WOLKOFF, A.W. (1973) Rate of evaporation of low solubility contaminants from water bodies to atmosphere. Environ. Sci. Technol., 7: 611-614.

MALYSCHEVA, M.V. (1988) [The effect of the skin route of administration of methyl isobutyl ketone on its toxicity.] Gig. i Sanit., 10: 79-80 (in Russian).

MICROBIOLOGICAL ASSOCIATES (1986) Subchronic toxicity of methyl isobutyl ketone in Sprague-Dawley rats. Preliminary report, Research Triangle Park, North Carolina, Research Triangle Park Institute, (Study No. 5221.04).

MITI (1978) The biodegradability and bioaccumulation of new and existing chemical substances, Tokyo, Ministry of International Trade and Industry, Chemical Products Safety Division, Basic Industries Bureau.

MOSHLAKOVA, L.A., & INDINA, T.V. (1986) [Gas chromatographic determination of ketones present simultaneously in the air of the work area and in the washings from workers' skin. Gig. i Sanit., 2: 90-91 (in Russian).

NAGANO, M., HARADA, K., MISUMI, J., & NOMURA, S. (1988) [Effect of methyl isobutyl ketone on methyl n-butyl ketone neurotoxicity in rats.] Sangyo Igaku, 30: 50-51 (in Japanese).

NIOSH (1984) NIOSH manual of analytical methods: ketones I, 3rd ed., Cincinnati, Ohio, US National Institute for Occupational Safety and Health,Vol. 2 (No. 1300).

NTIS (1985) Scientific literature review of aliphatic ketones, secondary alcohols and related esters in flavour usage. Volume 1 - Introduction and summary, tables of data bibliography: Al, Washington, DC, National Technical Information Service, Part 2 (PB85-141059).

O'DONOGHUE, J.L., HAWORTH, S.R., CURREN, R.D., KIRBY, P.E., LAWLOR, T., MORAN, E.J., PHILLIPS, R.D., PUTNAM, D.L., ROGERS-BACK, A.M., SLESINSKI, R.S., & THILAGAR, A. (1988) Mutagenicity studies on ketone solvents: methyl ethyl ketone, methyl isobutyl ketone and isophorone. Mutat. Res., 206: 149-161.

OECD (1977) The asessment of environmental chemicals: production figures and use patterns for some high volume chemicals, Paris, Organization for Economic Cooperation and Development (ENV/Chem./77.6).

OECD (1984) Data interpretation guides for initial hazard assessment of chemicals (provisional), Paris, Organisation for Economic Cooperation and Development, p. 31.

OH, S.J. & KIM, J.M. (1976) Giant axonal swelling in "Huffer's" neuropathy. Arch. Neurol., 33: 583-586.

PANSON, R.D. & WINEK, C.L. (1980) Aspiration toxicity of ketones. Clin. Toxicol., 17: 271-317.

PELLIZZARI, E.D., HARTWELL, T.D., HARRIS, B.S.H., WADDELL, R.D., WHITAKER, D.A., & ERICKSON, M.D. (1982) Purgeable organic compounds in mothers' milk. Bull. environ. Contam. Toxicol., 28: 322-328.

PHILLIPS, R.D., MORAN, E.J., DODD, D.E., FOWLER, E.H., KARY, C.D., &O'DONOGHUE, J. (1987) A 14-week vapor inhalation toxicity study of methyl isobutyl ketone. Fundam. appl. Toxicol., 9: 380-388.

PILON, D. (1987) Interaction cétones/hydrocarbures halogènes: Utilization des métabolites cétoniques comme indices d'exposition aux cétones, Montreal, University of Montreal (Ph.D. Thesis).

PILON, D., BRODEUR, J., & PLAA, G.L. (1988) Potentiation of carbon tetrachloride-induced liver injury by ketonic and ketogenic compounds: role of the CCl_4 dose. Toxicol. appl. Pharmacol., 94: 183-190.

PLAA, G.L. & AYOTTE, D. (1985) Taurolithocholate-induced intrahepatic cholestasis: potentiation by methyl isobutyl ketone and methyl n-butylketone in rats. Toxicol. appl. Pharmacol., 80: 228-234.

PRICE, K.S., WAGGY, G.T., & CONWAY, R.A. (1974) Brine shrimps bioassay and seawater BOD of petrochemicals. J. Water Pollut. Control Fed., 46: 63-77

RACCIO, J.M. & WIDOMSKI, J.R. (1981) Quality control of flavors in soft drinks and the analysis of residual solvents in food packaging films utilizing headspace sampling with open tubular columns. Chromatogr. Newsl., 9(2): 42-45.

RIPPSTEIN, W.J. & COLEMAN, M.E. (1984) [Toxicological evaluation on the Columbian spacecraft.] Kosmet. Biol. Aviakosm. Med., 18: 87-96 (in Russian).

RTECS (1987) Registry of toxic effects of chemical substances, 1985-86 ed., Cincinnati, Ohio, National Institute for Occupational Safety and Health, Vol. 1- 6 (DHSS (NIOSH) Publication No. 87-114).

RUTH, J.H. (1986) Odor threshold and irritation levels of several chemical substances: a review. Am. Ind. Hyg. Assoc. J., 47: A142-A151.

SABROE, S. & OLSEN, J. (1979) Health complaints and work conditions among lacquerers in the Danish furniture industry. Scand. J. soc. Med., 7: 97-104

SATO, A. & NAKAJIMA, T. (1979) Partition coefficients of some aromatic hydrocarbons and ketones in water, blood and oil. Brit. J. ind. Med., 36, 231-234.

SAWHNEY, B.L. & KOZLOSKI, R.P. (1984) Organic pollutants in leachates from landfill sites. J. environ. Qual., 13: 349-352.

SAX, N.I. (1979) Dangerous properties of industrial materials, 6th ed, New York, Van Norstrand Reinhold Company, p. 750.

SELKOE, D.J., LUCKENBILL-EDDS, L., & SHELANSKI, M.L. (1978) Effects of neurotoxic industrial solvents on cultured neuroblastoma cells: methyl n-butyl ketone, n-hexane, and derivatives. J. Neuropathol. exp. Neurol., 37: 768-789.

SHELL (1957) Methyl isobutyl ketone: industrial hygiene bulletin, New York, Shell Chemical Corporation, pp. 57-112.

SILVERMAN, L., SCHULTE, H.F., & FIRST, M.W. (1946) Further studies on sensory response to certain industrial solvent vapors. J. ind. Hyg. Toxicol., 28: 262-266.

SMYTH, H.F. (1956) Hygienic standards for daily inhalation. Am. Ind. Hyg. Assoc. J., 17: 129-266.

SMYTH, H.F., CARPENTER, C.P., & WEIL, C.S. (1951) Range-finding toxicity data: list IV. Arch. ind. Hyg. occup. Med., 4: 119-122.

SPECHT, H. (1938) Acute response of guinea pigs to inhalation of methyl isobutyl ketone, Washington, DC, US Public Health Service, pp. 292-300 (US Public Health Report No. 53).

SPECHT, H., MILLER, J.W., VALAER, P.J., & SAYERS, R.R. (1940) Acute response of guinea pigs to the inhalation of ketone vapours, Washington, DC, US Public Health Service, Division of Industrial Hygiene (NIH Bulletin No. 176).

SPENCER, P.S. & SCHAUMBURG, H.H. (1976) Feline nervous system response to chronic intoxication with commercial grades of methyl n-butyl ketone, methyl isobutyl ketone and methyl ethyl ketone. Toxicol. appl. Pharmacol., 37: 301-311.

SPENCER, P.S., SCHAUMBURG, H.H., RALEIGH, R.L., & TERHAAR, C.J. (1975) Nervous system degeneration produced by the industrial solvent methyl n-butyl ketone. Arch. Neurol., 32: 219-222.

TNO (1983a) Volatile compounds in food: Quantitative data, Zeist, Netherlands, Organization for Applied Scientific Research, Vol. 2.

TNO (1983b) Volatile compounds in food: Qualitative data, Zeist, Netherlands, Organization for Applied Scientific Research.

TNO (1986) Volatile compounds in food: Quantitative data, Zeist, Netherlands, Organization for Applied Scientific Research, Vol. 5.

TNO (1987) Volatile compounds in food: Supplement 4, Zeist, Netherlands, Organization for Applied Scientific Research.

TOMCZYK, H. & ROGACZEWSKA, T. (1979) [Gas chromatographic determination of airborne methyl isobutyl ketone, methyl isobutyl carbinol, acetone, toluene and o-xylene.] Med. Pr., **XXX**(6): 417-423 (in Polish).

TYL, R.W. (1984) A teratologic evaluation of methyl isobutyl ketone in Fischer 344 rats and CD-1 mice following inhalation exposure, Washington, DC, Chemical Manufacturers Association (Bushy Run Research Center Report No. 47.505).

TYL, R.W., FRANCE, K.A., FISHER, L.C., PRITTS, I.M., TYLER, T.R., PHILLIPS, R.D., & MORAN, E.J. (1987) Developmental toxicity evaluation of inhaled methyl isobutyl ketone in Fischer 344 rats and CD-1 mice. Fundam. appl. Toxicol.,**8**: 319-327.

VERNOT, E.H., MACEWEN, J.D., & HARRIS, E.S. (1971) Continuous exposure of animals to methyl isobutyl ketone, Ohio, Wright Patterson AFB, Aerospace Medical Research Laboratory (US NTIS AD Report No. 751443).

VERSCHUEREN, K. (1983) Handbook of environmental data on organic chemicals, 2nd ed., New York, Van Nostrand Reinhold Company, pp. 459-461.

VEZINA, M. & PLAA, G.L. (1987) Potentiation by methyl isobutyl ketone of the cholestasis induced in rats by a manganese-bilirubin combination or manganese alone. Toxicol. appl. Pharmacol. 91: 477-483.

VEZINA, M. & PLAA, G.L. (1988) Methyl isobutyl ketone metabolites and potentiation of the cholestasis induced in rats by a manganese-bilirubin combination or manganese alone. Toxicol. appl. Pharmacol. 92: 419-427.

VEZINA, M., AYOTTE, P., & PLAA, G.L. (1985) Potentiation of necrogenic and cholestatic liver injury by 4-methyl-2-pentanone. Can. Fed. Biol. Soc., **28**: 221.

WEBB, R.G., GARRISON, A.W., KEITH, L.H., & MCGUIRE, J.H. (1973) Current practice in GC-MS analysis of organics in water, Washington, DC, US Environmental Protection Agency (EPA Report No. R2-73-277) (NTIS PB 224 947/2).

WELLER, J.P. & WOLF, M. (1989) Mass spectroscopy and headspace gas chromatography. Beitr. gerichtl. Med., **47**: 525-532.

ZAKHARI, S., LEVY, P., LIEBOWITZ, M., & AVIADO, D.M. (1977) Acute oral, intraperitoneal, and inhalation toxicity of methyl isobutyl ketone in the mouse. In: Goldberg, L., ed. Isopropanol and ketones in the environment, Cleveland, Ohio, CRC Press, Part 3, Chapter 10-14, pp. 93-133.

ZLATKIS, A. & LIEBICH, H.M. (1971) Profile of volatile metabolites in human urine. Clin. Chem., 17: 592-594.

RESUME

La méthylisobutylcétone est un liquide limpide d'odeur douceâtre produite en vue d'une vaste utilisation commerciale comme solvant. On peut la doser par chromatographie en phase gazeuse avec détection par ionisation de flamme. Elle s'évapore rapidement dans l'atmosphère où elle subit une photoconversion à brève échéance. La méthylisobutylcétone est facilement biodégradable et, compte tenu de sa solubilité moyenne dans l'eau et de son faible coefficient de partage entre l'octanol et l'eau, elle devrait présenter un faible potentiel de bioaccumulation. Les limites d'exposition professionnelle sont de 100-400 mg/m³ (moyenne pondérée par rapport au temps: TWA) et de 5-300 mg/m³ (valeur plafond: CLV) selon les pays.

La méthylisobutylcétone est rapidement métabolisée en produits d'excrétion solubles dans l'eau et sa toxicité aiguë générale est faible chez l'animal après exposition par voie orale ou respiratoire. L'expérimentation animale n'a pas révélé d'axonopathie périphérique. On ne dispose pas de données précises sur la CL_{50}. Une exposition de 4 heures à une concentration de 16 400 mg/m³ (4000 ppm) a été mortelle pour des rats. La méthylisobutylcétone liquide ou sous forme de vapeurs à la concentration de 10 à 410 mg/m³ (2,4 à 100 ppm) est irritante pour les yeux et les voies respiratoires supérieures. Des concentrations allant jusqu'à 200 mg/m³ (50 ppm) n'ont produit aucun effet sensible sur l'homme lors d'épreuves portant sur le temps de réaction et le calcul mental. Un contact prolongé ou répété avec la peau peut produire un dessèchement et des crevasses. L'aspiration accidentelle de méthylisobutylcétone liquide peut provoquer une pneumonie chimique.

Lors d'une étude de 90 jours effectuée en gavant des rats, on a obtenu une dose sans effet observable de 50 mg/kg par jour. Des études d'inhalation de 90 jours portant sur des rats et des souris à des concentrations allant jusqu'à 4100 mg/m³ (1000 ppm) n'ont pas révélé de signes de toxicité engageant le pronostic vital. Toutefois des altérations morphologiques réversibles liées à l'administration de ce composé ont été observées au niveau du foie et des reins. Dans un certain nombre d'études, on a observé une hypertrophie du foie dès la dose de 1025 mg/m^3 (250 ppm). Exposés à 4100 mg/m^3 (1000 ppm) pendant 50 jours, des poulets ont présenté une induction des enzymes microsomiques. A doses plus élevées (jusqu'à 8180 mg/m^3, 1996 ppm) les effets se limitaient à un accroissement du poids du foie sans lésion histologique. Lors d'études de 90 jours effectuées sur des souris, des rats, des chiens et des singes, seuls les rats mâles ont présenté des altérations histologiques : présence de gouttelettes hyalines dans les tubules proximaux des reins (néphrose tubulaire toxique à inclusions hyalines). Cet effet s'est révélé réversible et sa portée en toxicologie humaine demeure incertaine. Il est possible que la potentialisation de la toxicité des alcanes halogénés par la méthylisobutylcétone repose sur une induction enzymatique. La méthylisobutylcétone potentialise également l'effet cholestatique du manganèse administré avec ou sans bilirubine.

Des babouins exposés pendant sept jours à une dose de 205 mg/m^3 (50 ppm) ont présenté des troubles neuro-comportementaux.

La méthylisobutylcétone est foetotoxique à une concentration manifestement toxique pour la mère (12 300 mg par m^3, 3000 ppm) mais elle n'est pas embryotoxique ni tératogène à cette concentration. A la concentration de 4100 mg/m^3 (1000 ppm), on n'a pas observé d'embryotoxicité, de foetotoxicité ni de tératogénicité chez les rats et les souris.

On a recherché la génotoxicité éventuelle de la méthylisobutylcétone en pratiquant un certain nombre d'épreuves à court terme, consistant notamment en tests sur des cellules mammaliennes, des bactéries et des levures ainsi que dans la recherche de micro-noyaux chez la souris. Les résultats indiquent que la méthylisobutylcétone n'est pas génotoxique. En ce qui concerne la génotoxicité à long terme ou la cancérogénicité, on ne dispose d'aucune donnée.

A la concentration de 410 mg/m^3 (100 ppm), la méthylisobutylcétone peut produire des symptômes chez l'homme consistant en irritation occulaire, migraines,

Résumé

nausées, vertiges et fatigue et qui correspondent à un effet dépresseur réversible sur le système nerveux central. Toutefois, rien n'indique l'existence de lésions permanentes.

La toxicité pour les organismes et micro-organismes aquatiques est faible.

Compte tenu de la volatilité relativement forte de la méthylisobutylcétone, de sa photoconversion rapide dans l'atmosphère, de sa biodégradabilité et de sa faible toxicité pour les mammifères et la faune aquatique, il est vraisemblable que cette substance ne peut exercer d'effets néfastes sur l'environnement qu'à la suite de déversements accidentels ou de la décharge incontrôlée d'effluents industriels.

EVALUATION DES RISQUES POUR LA SANTE HUMAINE ET DES EFFETS SUR L'ENVIRONNEMENT

1. Evaluation des effets sur l'environnement

La méthylisobutylcétone ne devrait pas persister dans l'environnement. Elle se volatilise lentement à partir du sol et de l'eau et subit une biodégradation rapide dans l'eau douce et l'eau salée. Dans l'atmosphère, on pense qu'elle est décomposée par les radicaux libres OH avec une demi-vie d'environ 14 heures. Elle ne s'accumule probablement pas et présente une faible toxicité pour les micro-organismes, les poissons, les algues et les invertébrés aquatiques. Ce n'est qu'en cas de déversement accidentel ou de rejet incontrôlé de déchets que cette substance est susceptible d'atteindre des concentrations toxiques pour les êtres vivants.

2. Evaluation des risques pour la santé humaine

La population générale n'est exposée qu'à de faibles concentrations de méthylisobutylcétone. On en a décelé de faibles quantités dans les denrées alimentaires et dans l'eau de consommation ou autres boissons (produits panifiés, 10,9 mg/kg; produits laitiers, 11,5 mg/kg; gélatines, puddings, 10,9 mg/kg; boissons diverses, 10,2 mg/kg). En ce qui concerne la population générale, deux pays ont fixé des concentrations maximales dans l'air ambiant qui se situent dans les limites de 0,1 à 0,2 mg/m^3.

L'exposition professionnelle se produit notamment lors de la production et de l'utilisation de vernis, de peintures et de solvants d'extraction. La principale voie de pénétration est l'inhalation. Le faible seuil olfactif (1,64 mg/m^3) et les effets irritants de cette substance peuvent avertir de la présence de fortes concentrations. L'exposition à ces concentrations de 10 à 410 mg/m^3 (2,4 à 100 ppm) a produit une irritation perceptible au niveau des yeux, du nez et de la gorge; à 820 mg/m^3 (200 ppm), on éprouve une sensation de malaise. Entre 10 et 410 mg par m^3 (2,4 à 100 ppm), on a également observé des céphalées, des nausées et des vertiges. Une exposition de deux heures à des concentrations allant jusqu'à 200 mg/m³ (50 ppm) n'a pas produit d'effets sensibles à en juger par un simple test de temps de réaction et de calcul mental.

On ne dispose que d'un seul rapport faisant état d'une exposition professionnelle de longue durée à une dose de 2050 mg/m³ (500 ppm), 20 à 30 minutes par jour et à 328 mg/m³ (80 ppm) pour la majeure partie du reste de la journée. Plus de la moitié des 19 ouvriers exposés se sont plaints de faiblesse, de perte d'appétit, de maux de tête, d'irritation oculaire, de maux d'estomac, de nausées, de vomissements et de maux de gorge. Quelques ouvriers ont éprouvé de l'insomnie, de la somnolence et une perte d'équilibre. Chez quatre d'entre eux on a constaté une légère hypertrophie du foie et chez six autres une colite non spécifique. Cinq années plus tard, les méthodes de travail s'étaient considérablement améliorées et les concentrations maximales réduites au cinquième des teneurs précédentes. Quelques ouvriers se sont encore plaints d'irritation au niveau des yeux et des voies respiratoires supérieures ainsi que de symptômes digestifs et neurologiques. La contact prolongé avec la peau a provoqué une irritation cutanée et des crevasses.

Il ressort de l'expérimentation animale que la toxicité générale aiguë de la méthylisobutylcétone est faible, par voie orale ou par inhalation. Lors d'une étude de 90 jours, des rats Sprague-Dawley ont reçu par gavage de la méthylisobutylcétone à des doses quotidiennes de 50, de 250 et 1000 mg/kg de poids corporel. On a noté une léthargie chez des animaux du groupe soumis à la dose la plus élevée et, chez les mâles, une réduction du gain de poids. Les animaux de ce groupe présentaient une néphropathie généralisée, avec augmentation du poids relatif des reins et une hypertrophie du foie. Cette augmentation du poids relatif des reins était également notable chez les animaux soumis à la dose de 250 mg/kg, mais à cette dose, on ne constatait qu'une légère hypertrophie du foie chez les mâles. Quelle que soit la dose, on n'a pas constaté de lésion histopathologique au niveau du foie ou des autres tissus. La dose sans effet observable a été évaluée à 60 mg/kg et par jour. Lors d'une étude de même durée au cours de laquelle on a fait inhaler à des rats et des souris des concentrations allant jusqu'à 4100 mg/m³ (1000 ppm) on n'a pas relevé de signes de toxicité engageant le pronostic vital. Toutefois des altérations morphologiques réversibles liées à l'administration de cette substance ont été observées au niveau du foie et des reins. A la concentration de 4100 mg par m³, on observait des signes de dépression du système nerveux central. La méthylisobutylcétone a provoqué une augmentation du poids du foie (à une dose supérieure à 1025 mg/m³, soit 250 ppm) et provoqué l'induction des enzymes microsomiques du foie. C'est ce dernier mécanisme qui serait à la base de l'exacerbation de la toxicité des alcanes halogénés et de la potentialisation de la neurotoxicité du n-hexane. Lors d'études de 90 jours sur des souris, des rats, des chiens et des singes, on a observé, chez les rats seulement, l'apparition d'inclusions hyalines dans les tubules proximaux des reins (néphrose tubulaire à inclusions hyalines d'origine toxique). Cet effet observé chez les rats mâles était réversible et il est douteux qu'il ait une signification quelconque en toxicologie humaine. La méthylisobutylcétone réduit l'activité de l'alcool-déshydrogénase hépatique chez la souris *in vitro*. On a également constaté qu'elle potentialisait les effets cholestatiques du manganèse en présence ou en l'absence de bilirubine.

Des rats et des souris exposés par inhalation à des concentrations de 1230, 4100 ou 12 300 mg/m³ (300, 1000 ou 3000 ppm) du sixième au quinzième jours de la gestation puis sacrifiés le vingt-et-unième jour (rats) ou le dix-huitième jour (souris), ont présenté des signes marqués d'intoxication à la concentration la plus forte. Cette concentration était foetotoxique (réduction du poids foetal et ossification retardée) mais n'était ni embryotoxique ni tératogène. Aux concentrations de 4100 et 1230 mg/m³, on n'a constaté aucune toxicité pour les mères ni signe d'embryotoxicité, de foetotoxicité ou de tératogénicité.

La méthylisobutylcétone n'a pas produit de mutation génique dans des systèmes d'épreuve bactériens (Salmonella typhimurium et *Escherichia coli*), qu'il y ait ou non activation métabolique. On a également obtenu des résultats négatifs dans différentes épreuves (avec ou sans activation métabolique) à la recherche de conversions géniques mitotiques dans des levures *(Saccharomyces cerevisiae)* ou lors d'épreuves de mutation génique sur des cellules mammaliennes en culture (lymphome murin). La recherche *in vitro* d'une synthèse anarchique de l'ADN dans des hépatocytes primaires de rat et de lésions chromosomiques structurales dans des cellules de foie de rat en culture (RL4) s'est révélée négative. Chez la souris, la recherche *in vivo* de micro-noyaux s'est également révélée négative. Toutes ces données montrent que la méthyl-isobutylcétone n'est pas génotoxique.

RECOMMANDATIONS

Les concentrations de méthylisobutylcétone auxquelles la population, dans son ensemble, est susceptible d'être exposée, ne présentent vraisemblablement aucun danger. La principale voie d'exposition professionnelle est la voie respiratoire, aussi les concentrations atmosphériques devront être maintenues en dessous des limites recommandées d'exposition professionnelle, grâce à un aménagement convenable des procédés de production et à des moyens mécaniques tels que la ventilation. Il convient d'éviter toute contamination de la peau et des yeux. Des vêtements protecteurs appropriés ainsi que des masques respiratoires doivent être placés dans les ateliers confinés; ils seront utilisés en cas d'urgence ou pour effectuer certaines opérations d'entretien. La méthylisobutylcétone est inflammable et doit donc être manipulée avec les précautions d'usage.

La méthylisobutylcétone présente une faible toxicité pour les micro-organismes et pour les poissons et sa demi-vie dans l'environnement est courte. Il s'ensuit qu'elle ne présente aucun risque pour l'environnement, dans la mesure où des mesures appropriées sont prises pour réduire les émissions au minimum. La décharge de quantités importantes dans l'environnement pourrait avoir localement des effets indésirables.

RECHERCHES A EFFECTUER

1. La méthylisobutylcétone affecte un certain nombre de systèmes enzymatiques. Elle peut donc avoir une influence sensible sur la biotransformation des produits xénobiotiques métabolisés par ces enzymes. Etant donné que l'homme est généralement exposé à plusieurs composés différents, il faudrait entreprendre des études sur les effets combinés de mélanges contenant de la méthylisobutylcétone.

2. On ne dispose que de très peu d'informations sur la relation dose-réponse relative aux effets toxiques de la méthylisobutylcétone sur le système nerveux central (par exemple temps de réaction, effets comportementaux), sur les voies respiratoires supérieures, sur les muqueuses et sur la fonction rénale. Il faudrait obtenir davantage de renseignements sur la toxicocinétique de cette cétone, soit seule soit associée à d'autres solvants. Il faudra également étudier la pénétration percutanée de la méthylisobutylcétone.

3. Il faudrait entreprendre des études épidémiologiques pour élucider les effets exercés à long terme sur le système nerveux central par des concentrations moyennes de méthylisobutylcétone soit seule soit associée à d'autres solvants.

RESUMEN

La metil isobutil acetona (MIBA) es un líquido transparente de buen olor que se produce a escala comercial y tiene un uso muy extendido como disolvente. Puede medirse mediante cromatografía de gases con detección de ionización de llama. Se evapora rápidamente a la atmósfera, donde se fototransforma en poco tiempo. La MIBA es fácilmente biodegradable, lo que, junto con su moderada solubilidad en el agua y su reducido coeficiente de partición octanol/agua, sugiere que tiene un bajo potencial de bioacumulación. Los límites de exposición profesional varían entre 100-410 mg/m^3 (promedio ponderado en el tiempo) y 5-300 mg/m^3 (valor máximo) en distintos países.

La MIBA se metaboliza fácilmente para dar productos de excreción hidrosolubles y su toxicidad sistémica aguda en animales es baja por las vías de exposición oral y de inhalación. No se ha observado axonopatía periférica en estudios realizados en animales. No se dispone de datos exactos sobre la CL$_{50}$. La exposición a 16 400 mg/m^3 (4000 ppm) durante 4 horas resultó letal para la rata. Las concentraciones de MIBA líquida y en vapor comprendidas entre 10 y 410 mg/m^3 (2,4-100 ppm) son irritantes para los ojos y las vías respiratorias superiores. Con concentraciones de hasta 200 mg/m^3 (50 ppm) no se observaron en el hombre efectos significativos en una prueba sencilla de tiempo de reacción ni en una prueba de aritmética mental. El contacto cutáneo prolongado o repetido puede desecar y descamar la piel. La aspiración accidental de MIBA líquida puede provocar pneumonitis química.

En un estudio de ceba de ratas durante 90 días, se determinó un nivel sin efecto observado de 50 mg/kg. En estudios de inhalación durante 90 días en ratas y ratones, concentraciones de hasta 4100 mg/m^3 (1000 ppm) no produjeron ningún signo de toxicidad con peligro para la vida. No obstante, se notificaron cambios morfológicos reversibles relacionados con el compuesto en el hígado y el riñón. En varios estudios se observó que concentraciones de MIBA tan bajas como 1025 mg/m^3 (250 ppm) eran capaces de aumentar el tamaño del hígado. Mediante la exposición a 4100 mg/m^3 (1000 ppm) durante 50 días, se indujo actividad metabólica en las enzimas microsómicas del hígado del

Resumen

pollo. A dosis más elevadas (hasta 8180 mg/m^3, 1996 ppm) los efectos se limitaron a un aumento del peso del hígado sin lesiones histológicas. En estudios durante 90 días con ratones, ratas, perros y monos, sólo las ratas macho presentaron corpúsculos hialinos en los túbulos proximales del riñón (nefrosis tubulotóxica de corpúsculos hialinos). Este efecto en la rata macho resultó ser reversible y de dudosa importancia para el hombre. La inducción enzimática puede ser la base de la potenciación de la toxicidad de los haloalcanos por la MIBA. Se observó también que la MIBA era capaz de potenciar el efecto colestático del manganeso administrado con o sin bilirrubina.

En papiones expuestos durante 7 días a 205 mg/m^3 (50 ppm), se observaron efectos en el neurocomportamiento.

La MIBA resulta fetotóxica a una concentración que produce sin lugar a dudas toxicidad materna (12 300 mg/m^3, 3000 ppm) pero no es embriotóxica ni teratogénica en esa concentración. En una concentración de 4100 mg/m^3 (1000 ppm), no resultó ni embriotóxica, ni fetotóxica ni teratogénica en la rata ni en el ratón.

Se ha estudiado la genotoxicidad de la MIBA en varios ensayos a corto plazo, inclusive pruebas in vitro con bacterias, levaduras y células de mamíferos y un ensayo de micronúcleos en el ratón. Esos estudios indican que la MIBA no es genotóxica. No se dispone de informes sobre estudios a largo plazo ni estudios de carcinogenicidad.

Aunque con una concentración de 410 mg/m^3 (100 ppm) la MIBA puede inducir en el hombre síntomas como irritación ocular, dolores de cabeza, náuseas, mareos y fatiga, que corresponden a un efecto reversible de depresión del sistema nervioso central, no existen pruebas de que produzca lesiones permanentes en el sistema nervioso.

La toxicidad de la MIBA para organismos y microorganismos acuáticos es baja.

La volatilidad relativamente alta de la MIBA, su rápida fototransformación en la atmósfera, su fácil biodegradación y su baja toxicidad para mamíferos y organismos acuáticos indican que los efectos medioambientales adversos de esta sustancia probablemente sólo se producirán como consecuencia de vertidos accidentales o de efluentes industriales no controlados.

EVALUACION DE LOS RIESGOS PARA LA SALUD HUMANA Y DE LOS EFECTOS EN EL MEDIO AMBIENTE

1. **Evaluación de los efectos en el medio ambiente**

   La MIBA tiene pocas probabilidades de persistir en el medio ambiente. Se volatiliza poco a poco desde el suelo y el agua y se biodegrada fácilmente en agua dulce y salada. En la atmósfera, se ha calculado que la MIBA es degradada por los radicales OH· con una semivida de aproximadamente 14 horas. En principio, la MIBA no se bioacumula y tiene una toxicidad baja para micro-organismos, peces, algas e invertebrados acuáticos. Sólo en los casos de vertido accidental o de evacuación inadecuada de desechos en el medio ambiente es probable que los niveles de MIBA provoquen toxicidad en los organismos del entorno.

2. **Evaluación de los riesgos para la salud humana**

   La población general está expuesta a niveles reducidos de MIBA. Se han detectado sólo pequeñas cantidades en los alimentos, el agua potable y otras bebidas (alimentos horneados, 10,9 mg/kg; productos lácteos congelados, 11,5 mg/kg; gelatinas y budines, 10,9 mg/kg; bebidas, 10,2 mg/kg). En cuanto a la exposición de la población general, dos países han definido concentraciones máximas en el aire entre 0,1 y 0,2 mg/m^3.

   La exposición profesional tiene lugar especialmente en la producción y utilización de lacas, pinturas y disolventes de extracción. La principal vía de entrada es por inhalación. El bajo umbral olfativo (1,64 mg/m^3) y los efectos irritantes pueden servir como indicadores de las concentraciones elevadas. La exposición a niveles de 10-410 mg/m^3 (2,4-100 ppm) produjo irritación perceptible de los ojos, la nariz, o la garganta y 820 mg/m^3 (200 ppm) produjeron molestias. Con un nivel de 10-410 mg/m^3 (2,4-100 ppm) también se produjeron síntomas como dolor de cabeza, náuseas y vértigos. No se observaron efectos significativos debidos a una exposición durante 2 horas a hasta 200 mg/m^3 (50 ppm) al realizar una prueba sencilla de tiempo de reacción ni una prueba de aritmética mental.

En el único informe sobre un estudio de la exposición profesional a largo plazo, en el que se expuso a trabajadores a 2050 mg MIBA/m^3 (500 ppm) durante 20-30 minutos al día y a 328 mg/m^3 (80 ppm) durante la mayor parte del resto de la jornada laboral, más de la mitad de los 19 trabajadores se quejaron de debilidad, pérdida del apetito, dolores de cabeza, irritación ocular, dolor de estómago, náuseas, vómitos y dolor de garganta. Algunos trabajadores sufrieron insomio, somnolencia y cierta inestabilidad. En cuatro se observó un ligero agrandamiento del hígado y en seis se observó colitis no específica. Al cabo de 5 años, las prácticas laborales habían mejorado en gran medida y las concentraciones más elevadas se redujeron a aproximadamente la quinta parte del nivel anterior. Algunos trabajadores siguieron quejándose de irritación en los ojos y de las vías respiratorias superiores así como de síntomas gastrointestinales y del sistema nervioso central. El contacto cutáneo prolongado con la MIBA provocó irritación y descamación de la piel.

En estudios en animales, la toxicidad sistémica aguda de la MIBA es baja por las vías oral y respiratoria. En un estudio de 90 días de duración, se cebó con MIBA a ratas Sprague-Dawley en dosis diarias de 50, 250 ó 1000 mg/kg de peso corporal. Se observó letargo en el grupo que recibió la dosis más alta y en los machos se observó una reducción del ritmo de aumento del peso corporal. En este grupo se observó nefropatía generalizada, con aumento del peso relativo del riñón y hepatomegalia. El peso relativo del riñón también aumentó en los animales alimentados con 250 mg/kg al día, y se observó ligera hepatomegalia sólo en los machos. No aparecieron lesiones histopatológicas en el hígado ni en otros tejidos con ninguna de las dosis administradas. Se concluyó que el nivel de efecto no observado era de 50 mg/kg al día. En estudios de inhalación durante 90 días realizados en ratas y ratones, las concentraciones de hasta 4100 mg/m^3 (1000 ppm) no originaron ningún signo de toxicidad que pusiera en peligro la vida. No obstante, se comunicó la observación de cambios morfológicos reversibles relacionados con el compuesto en el hígado y el riñón. Con niveles de 4100 mg/m^3 se observaron signos de depresión del sistema nervioso central. La MIBA fue capaz de aumentar el peso del hígado (concentración > 1025 mg/m^3 (250 ppm)) y de inducir el metabolismo microsómico hepático. Esto puede explicar la exacerbación de la toxicidad de los haloalcanos y la potenciación de la neurotoxicidad del n-hexano. En estudios de 90 días con ratones, ratas, perros y monos, sólo las ratas macho desarrollaron corpúsculos hialinos en los túbulos proximales del riñón (nefrosis tubulotóxica de corpúsculos hialinos). Este efecto en ratas macho resultó ser reversible y de dudosa importancia para el hombre. In vitro, la MIBA reduce la actividad de la deshidrogenasa alcohólica en el hígado del ratón. También se ha observado que potencia los efectos colestáticos del manganeso administrado con o sin bilirrubina.

En ratas y ratones expuestos a la inhalación de MIBA en concentraciones de 1230, 4100 ó 12 300 mg/m^3 (300, 1000 ó 3000 ppm) en los días 6 a 15 de la gestación y sacrificados el día 21 (ratas) o el día 18 (ratones), se observó una notable toxicidad materna a la concentración más elevada en ambas especies. Esta concentración produjo fetotoxicidad (peso del cuerpo fetal reducido y retraso en la osificación) pero no resultó embriotóxico ni teratogénico. A 4100 y 1230 mg/m^3 no se observó ni toxicidad materna ni síntomas de embriotoxicidad, fetotoxicidad o teratogenicidad.

La MIBA no indujo mutación genética en sistemas de ensayo bacterianos (Salmonella typhimurium y Escherichia coli), con o sin activación metabólica. También se obtuvieron resultados negativos en los ensayos (tanto con y sin activación metabólica) para la conversión mitótica de genes en levaduras (Saccharomyces cerevisiae) y en pruebas de mutación génica con cultivos de células de mamíferos (linfoma de ratón). Los ensayos in vitro sobre síntesis no controlada de ADN en hepatocitos primarios de rata y sobre lesiones cromosómicas estructurales en cultivos de células hepáticas de rata (RL4) resultaron negativos. En el ratón, un ensayo de micronúcleo in vivo resultó negativo. Estos datos indican que la MIBA no es genotóxica.

RECOMENDACIONES

En los niveles de MIBA a que está expuesta la población humana general, es poco probable que se plantee riesgo alguno. En el medio laboral, donde la principal vía de exposición es por inhalación, los niveles atmosféricos deben mantenerse por debajo de los límites de exposición profesional recomendados mediante procesos de trabajo y controles de ingeniería, inclusive la ventilación, de diseño adecuado. Debe evitarse la contaminación de la piel y de los ojos. Debe facilitarse el uso de prendas protectoras adecuadas y de protección respiratoria en los espacios cerrados, en casos de emergencia y para ciertas operaciones de mantenimiento. La MIBA es inflamable y debe manipularse teniendo esta característica en cuenta.

La MIBA tiene baja toxicidad para los microorganismos y los peces, y su semivida en el medio ambiente es corta. Por consiguiente, no presenta riesgos para el medio ambiente siempre que se apliquen las medidas de control adecuadas para reducir al mínimo las emisiones. El vertido en gran escala podría ejercer efectos adversos en el medio ambiente a escala local.

OTRAS INVESTIGACIONES

1. LA MIBA afecta a varios sistemas enzimáticos. Por lo tanto, puede influir de modo significativo en la biotransformación de sustancias biológicas externas que son metabolizadas por estas enzimas. Como las personas suelen estar expuestas a más de un compuesto, deben llevarse a cabo estudios sobre los efectos combinados de muestras que contengan MIBA.

2. Se dispone de muy poca información sobre las relaciones dosis-respuesta en cuanto a los efectos de la MIBA en el sistema nervioso central humano (por ejemplo, tiempo de reacción, efectos conductuales) en las vías respiratorias superiores y las mucosas, o en la función renal. Se necesita más información sobre la toxicocinética de la MIBA por sí sola y mezclada con otros disolventes. Debe evaluarse la penetración cutánea de la MIBA.

3. Deben emprenderse estudios epidemiológicos para dilucidar los efectos que tiene en el sistema nervioso la exposición a largo plazo a concentraciones moderadas de MIBA, por sí sola o mezclada con otros disolventes.

www.ingramcontent.com/pod-product-compliance
Ingram Content Group UK Ltd.
Pitfield, Milton Keynes, MK11 3LW, UK
UKHW021306180426
11947UKWH00015B/1055

9 789241 571173